# Calculus for the Forgetful

# Calculus for the Forgetful

**How to understand more and memorize less**

**Wojciech K. Kosek**

MagiMath Publishing, Colorado Springs

Wojciech K Kosek
Department of Mathematics and Computer Science
Colorado College

Edited by *Alison A. Anderson*
Published by *MagiMath Publishing*



MagiMath Publishing
1910 Vindicator Drive, Suite 105
Colorado Springs, CO 80919
USA

Library of Congress Control Number: 2007926476

ISBN 978-0-9795199-0-1

Printed in the United States of America
First printing

www.magimath.com

To Iwona and Magdalena

# Contents

# Preface

The purpose of this short book is to provide the reader with a concise treatment of single variable calculus. The main emphasis is on the understanding of ideas and facts, rather than their memorization. Informal, intuitive explanations are included for most facts, whenever feasible. Most of us tend to forget the fine details of proofs, but it is good to have a general understanding of why various statements are true. My goal is to help the reader understand the main ideas of calculus. There are certainly many topics which are omitted for the sake of brevity. This book does not aim to be an encyclopedic treatment of the subject. Instead, I try to stay with the core ideas, with just a few digressions here and there.

The book can be used as a supplemental text by students currently taking a calculus course. It is also for those who took calculus in the past and would like to organize their knowledge, for example, while taking another class in which calculus is needed. Those who plan to take the GRE could also benefit from reading the book. The size is small enough to be carried around to other classes in which calculus is used. There are more than 130 examples and many exercises in the text, and some more challenging problems at the end of each chapter. The problems are designed as an enhancement, but are not sufficient for most people to acquire proficiency in calculus.

I emphasize the core ideas and concepts, but include occasional digressions and comments which show possible avenues for further exploration. Many of these side trips are in smaller print, and skipping them will not cause the reader any difficulties in comprehending the rest. It is my hope that the reader will not feel intimidated by any of the nonessential material.

Everyone has a somewhat different learning style. It is my hope that this book will make calculus more enjoyable and easier to grasp for at least some readers. Chapter 1 contains a review of some basic facts and concepts used in calculus, including trigonometry and logarithms, which are a frequent source of headaches for young calculus apprentices. In Chapters 2 and 3 the concept of the derivative and methods of finding it are discussed. A limited selection of applications of derivatives is the topic of Chapter 4. The concept of the integral, methods of integration and some applications are discussed in Chapters 5, 6 and 7. The book concludes, in Chapter 8, with a brief introduction to infinite series.

In addition, there are four appendices, which provide some additional review of the background material (A); a summary of the book (D); answers to selected problems (C); and a few proofs, which are neat, but not essential for the general understanding of the subject (B).

I wish to thank all those who helped me make this book better, especially my students who were subjected to earlier versions of this work and who generously shared their ideas and thoughts with me. Special thanks go to Andrea Buchwald and Chris Kempes for their diligent proofreading and helpful comments. I am grateful to Courtney Gibbons for her help in making the figures and to Alison Anderson, Ph.D., for all the assistance in editing. I am also thankful to Professors Robin Wilson and John Watkins, for sharing their insights about book writing with me.

Observing all those who taught me mathematics and other subjects is how I developed my own teaching style. I owe a debt of gratitude to all of my teachers, including my parents; my high school teacher, Mrs. Maria Czerska-Lazarowicz, M.A.; my undergraduate advisors, Professors Roman Ger, Jerzy Klamka and Andrzej Świerniak; my Ph.D. advisor, Professor Isaac Kornfeld; and countless faculty and colleagues of mine, from whom I learned mathematics and how to explain it. I am grateful to my family for all their support.

Most of all, I wish to thank my wife Iwona and our daughter Magdalena, whose encouragement and infinite patience made this book possible.

Wojciech K Kosek, Colorado College

# Chapter 1

# Preliminary Concepts and Facts

## 1.1 Real numbers

The ancient Greeks discovered that the length of the diagonal of a square of side 1 is $\sqrt{2}$. This follows from what is known as the Pythagorean theorem: the sum of the areas of the squares built on the sides of a right triangle is the same as the area of the square built on the hypotenuse. This is clear from the picture: the area of the big square can be calculated in two ways: $(a+b)^2 = c^2 + 4 \cdot \frac{1}{2}ab$. Expanding the left-hand side gives

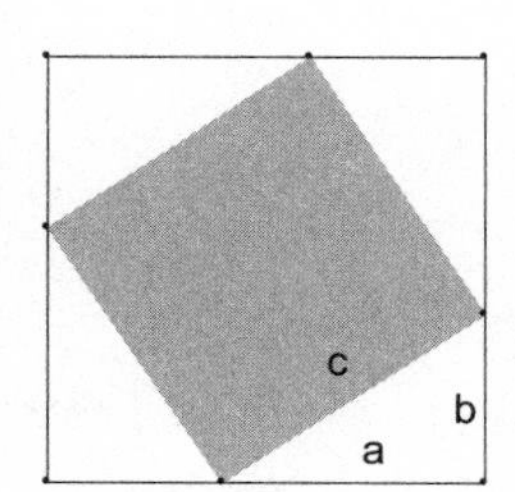


$$a^2 + b^2 = c^2.$$

What is the big deal about it? The Greeks believed that all numbers in nature can be represented as fractions of integers. After specifying two points on a straight line as 0 and 1, it is natural to identify every point on this line with a number: we call such numbers **real**. However, $\sqrt{2}$ *cannot* be represented as a ratio of two integers $\frac{m}{n}$ (see Appendix B for the proof). This was a serious problem: if mathematicians do not even know what a number is, what do they know? By now, we have come to terms with this realization: there are actually more real numbers which are not fractions of integers (irrationals) than there are those which can be written in that form. (We also realize that there are a lot more questions for which we do not have answers than those for which we do...)

The following is the standard classification of numbers, along with notation commonly used to denote them:

$\mathbb{N}$ : the natural numbers; whole positive integers and zero;

$\mathbb{Z}$ : the integers; positive, negative and zero;

$\mathbb{Q}$ : the rational numbers; fractions of integers;

$\mathbb{R}$ : all real numbers.

In modern times people came to appreciate an even larger set of numbers, the complex numbers $\mathbb{C}$, but we need to stay focused on calculus, and this issue is beyond our current interest.

Another common practice is the use of the interval notation: $(a, b]$ denotes the set of real numbers $x$, such that $a < x \leq b$; $(-\infty, a]$ stands for the set of real numbers $x$ such that $x \leq a$; and so on.

We use the symbol $|x|$ to denote the absolute value of $x$, which should be thought of as the distance from 0 to $x$. In a similar fashion, the distance between numbers $a$ and $x$ is $|x - a|$.

## 1.2 Coordinates on the plane

Just as there is a one-to-one correspondence between real numbers and the points on the line, we can also identify ordered pairs of real numbers with points on the plane. The distance between two points $P(x_1, y_1)$ and $Q(x_2, y_2)$ is calculated using the Pythagorean theorem:

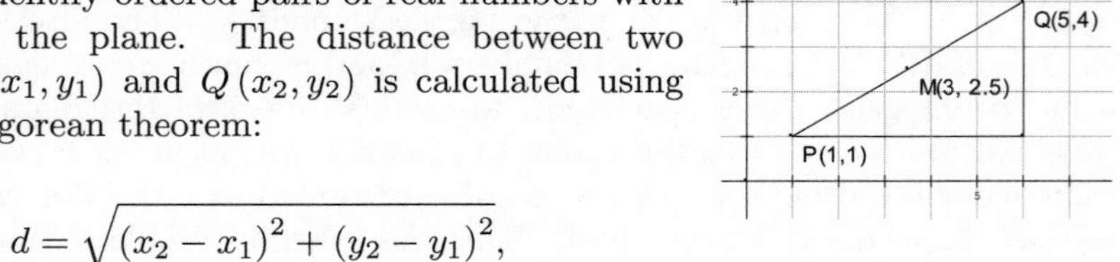


$$d = \sqrt{(x_2 - x_1)^2 + (y_2 - y_1)^2},$$

and the point in the middle between $P$ and $Q$ (the midpoint) is given by taking arithmetic averages of the coordinates:

$$M\left(\frac{x_1+x_2}{2},\frac{y_1+y_2}{2}\right).$$

The distance formula leads to the equation of a circle, centered at $(h.k)$, with radius $r$:

$$(x-h)^2+(y-k)^2=r^2.$$

## 1.3 Lines and linear functions

Let us consider a straight, nonvertical line, passing through two given points $P(x_1,y_1)$ and $Q(x_2,y_2)$. Given these two points, we can calculate a number

$$m=\frac{\Delta y}{\Delta x}=\frac{y_2-y_1}{x_2-x_1}. \tag{1.1}$$

It follows from the similarity of triangles that the value of the number $m$ does not depend on the choice of the points on the line; hence, for any other point $(x,y)$ on the line, we have

$$\frac{y-y_1}{x-x_1}=m. \tag{1.2}$$

Hence, to every straight, nonvertical line on the plane, we can associate a number $m$, called **slope**. The value of $m$ has a very tangible interpretation: it tells us the rate of change of $y$ with respect to $x$ (how steep the line is). Frequently, an equation of a line is written in a so-called "**point-slope**" form, which is obtained by multiplying equation (1.2) by $x-x_1$ and adding $y_1$ to both sides:

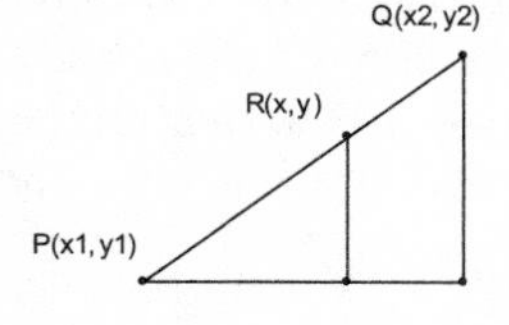


$$y=m(x-x_1)+y_1. \tag{1.3}$$

By multiplying out the expression on the right-hand side of the point-slope form and simplifying, we can get the **slope-intercept** form of the equation: $y=mx+b$, where the number $b$ is the $y$-intercept of the line. Both the slope-intercept and the point-slope forms have $y$ as a function of $x$, called a **linear function**. The concept of slope is crucial in calculus, as we will generalize it to the slope of a curve. We will call it the derivative. First, let us recall some of the facts about some common functions. For a quick review of necessary terminology, please check Appendix A.1.

## 1.4 Quadratic functions

The simplest function whose graph is not a straight line is a quadratic function of the form

$$f(x) = ax^2 + bx + c. \tag{1.4}$$

The graph of a quadratic function has a shape of a parabola, opening up if $a > 0$ and down if $a < 0$. It may cross the $x$-axis at two points, touch it at just one point, or stay on one side of it for all values of $x$. To focus our attention, let us assume for a moment that $a > 0$. If the function is written in the form $f(x) = a(x-h)^2 + k$, then the lowest value it can ever attain would be $f(h) = k$, since the expression $(x-h)^2$ is a square and the least it can be is zero. This would also make it easy to find the $x$-intercepts, which occur only if $k \leq 0$.

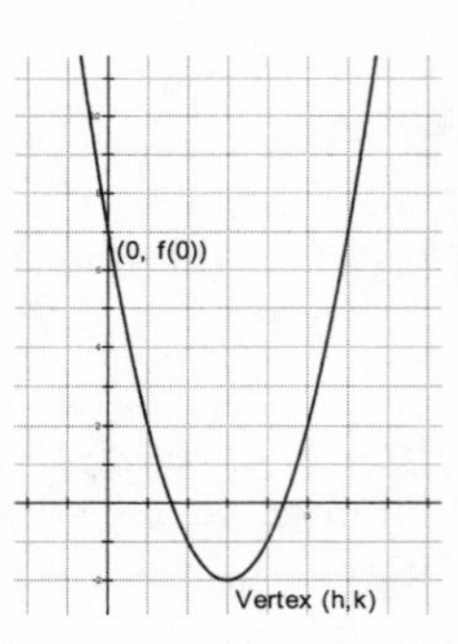


**Example 1** Let $f(x) = x^2 - 6x + 4$. To find the $x$-intercepts, we need to solve the equation $f(x) = 0$. Let us notice that $x^2 - 6x + 9$ is a square of $x - 3$. If we add 9 to both sides of the original equation, we obtain $x^2 - 6x + 9 + 4 = 9$. Subtracting 4 from both sides gives $(x-3)^2 = 5$. Thus, the $x$-intercepts are $3 \pm \sqrt{5}$. The number 9 used in this trick is calculated as $\left(\frac{6}{2}\right)^2$.

The procedure illustrated in the example is called "completing the square", and is used quite often in various settings. In general, this is how we do it:

$$\begin{aligned} f(x) &= ax^2+bx+c = a\left[x^2+\frac{b}{a}x\right]+c \\ &= a\left[x^2+\frac{b}{a}x+\left(\frac{b}{2a}\right)^2-\left(\frac{b}{2a}\right)^2\right]+c \qquad (1.5) \\ &= a\left(x+\frac{b}{2a}\right)^2-a\left(\frac{b}{2a}\right)^2+c \\ &= a\left(x+\frac{b}{2a}\right)^2-\frac{b^2-4ac}{4a}. \end{aligned}$$

Setting $f(x)=0$, we can now easily solve for

$$x=\frac{-b\pm\sqrt{b^2-4ac}}{2a}. \qquad (1.6)$$

The reader will likely agree that the quadratic formula for the solutions of the equation $ax^2+bx+c=0$ is worth remembering, unless we want to complete the square every time we need to solve a quadratic equation. Still, it is good to know how to do it. Also, it is clear from the quadratic formula (1.6) that the parabola has $x$-intercept(s) only if the expression under the radical (called the radicand) $b^2-4ac\geq 0$. The coordinates of the vertex $(h,k)$ are also easily found from the completed square form (1.5): $h=\frac{b}{2z}$ and $k=f(h)$, the value of the function at $x=h$.

## 1.5 Polynomials and rational functions

The next natural step is to allow higher powers of $x$ in the function. A **polynomial** is an expression of the form

$$P(x)=a_0+a_1x+a_2x^2+\cdots+a_nx^n.$$

The **degree** of a polynomial is the highest power to which the variable is raised.

So how do we find the $x$-intercepts, which are also called the **roots** of the polynomial? Formulas for the solutions of the equation of degree 3 and 4 were developed in Italy during the Renaissance. They are called Cardano (for the cubic) and Ferrari (for the quartic) formulas. We do not write them here, as they are quite complicated and are not used as often as the quadratic formula. However, computer algebra systems use them routinely. There are no formulas, in the traditional sense, for equations of degree 5 and higher!

The impossibility of finding solutions of the quintic in general (in terms of the usual four operations and taking roots) was shown by a young Norwegian mathematician, Niels Abel, in 1824. Just a few years later Abel described the exact conditions under which it is possible to express the roots in terms of the coefficients and the aforementioned operations. At about the same time, another brilliant young man, Evariste Galois, independently discovered the same. In short, Galois theory describes what kind of numbers one can get by applying the operations of addition, subtraction, multiplication, division and extraction of roots to the coefficients. The root of the equation of degree 5 or higher may be outside of this set of numbers. According to a legend, Galois spent the night before the duel in which he was killed, writing his mathematical theories. I do hope that this may inspire some of us to apply ourselves just a little harder.

Coming back to polynomial and rational functions: it is obvious that if $P(x) = (x-a) \cdot Q(x)$, for some other polynomial $Q$, then $P(a) = 0$. The converse is also true, but a little less obvious: if $P(a) = 0$, then $(x-a)$ is a factor of $P(x)$. This is a useful fact: if we happen to know that $a$ is a root of a polynomial, it must be possible to divide that polynomial by $x-a$, thus reducing its degree by one. It may then be easier to find other roots (see Problem 7). It also follows that a polynomial of degree $n$ can have at most $n$ different roots.

There is a lot more that can be said about these issues. Unfortunately, in most "real life" situations, finding the exact values of the roots of polynomials is not possible, and we have to settle for their numerical approximations.

A ratio of two polynomial functions is called a **rational** function. A problem frequently encountered in calculus is to find for what values of $x$ is some function positive. A simple method can be used to solve this problem for a rational function, provided that we can factor both the numerator and denominator.

**Example 2** Solve the inequality

$$\frac{x^2(x+3)}{(x-5)^3} \geq 0. \tag{1.7}$$

First, let us notice that the sign of either a product or a quotient of numbers does not depend on their actual values but only on their signs. The expression in inequality (1.7) has three "components": $x^2$, $x+3$ and $x-5$. Each component has a threshold value of $x$ for which it may change its sign. We need to keep track of the signs of each of the components and count how many of them are negative. Some people like to make a chart, but for me the quickest way is to draw a quick sketch. In this example, there are three threshold values: $-3$, $0$ and $5$, which in turn cut the real line into four intervals: $(-\infty,-3)$, $(-3,0)$, $(0,5)$ and $(5,\infty)$. The inequality is either true or false for all $x$ inside each of the intervals.

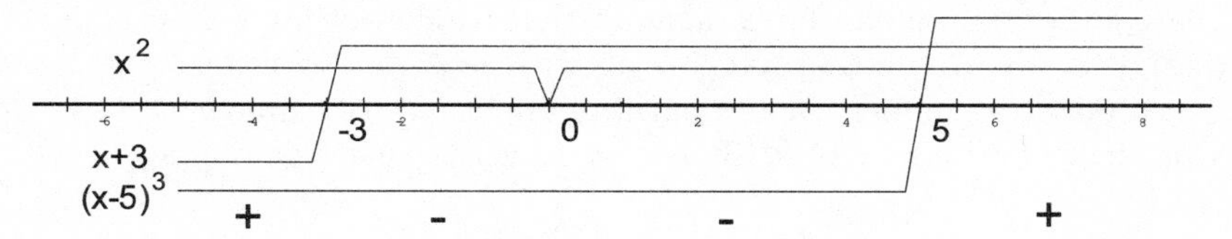


One way to indicate the sign of each expression is to sketch a line, above or below the $x$-axis. For example, $x+3$ gives a line below the $x$-axis for $x<-3$ and above it for $x>-3$. We do the same for $(x-5)^3$, since it has the same sign as $x-5$. The line corresponding to $x^2$ stays above the $x$-axis at all times, except for $x=0$. It is now easy to mark the sign of the whole rational expression in each of the intervals by counting the number of lines below the $x$-axis. Finally, we need to decide whether to include any of the threshold values in the solution. The key issue is their origin: the numbers which make the denominator zero can never be included, as the expression is undefined (in our case, we do not include 5). Since the inequality was not sharp ($>$), we should include $0$ and $-3$. Therefore, the solution set to inequality (1.7) is

$$(-\infty,-3] \cup \{0\} \cup (5,\infty).$$

In general, the roots of a rational function are those values of $x$, for which the numerator is equal to zero and the denominator is not. That is because if the denominator is zero, the fraction is undefined.

On the other hand, the zeroes of the denominator are candidates for vertical asymptotes. However, there may or may not be an asymptote at the value for which both numerator and denominator vanish. For example, consider a function $f(x) = \frac{(x-1)(x+2)}{x-1}$, which is simply $x+2$, as long as $x \neq 1$. A few examples of graphs of rational functions are provided in Problem 4.

## 1.6 Exponentials and logarithms

Exponential and logarithmic functions are of great importance in mathematics. Let us first review some facts about raising numbers to different exponents. While one can certainly raise negative numbers to integer powers, and to some fractional powers as well, we will restrict our attention to positive values of the base $b$. Let us pretend for a moment that we do not know how exponents work and we are attempting to come up with sensible definitions.

For a positive number $b$ and a natural number $n$, we define $b^n = b \cdot b \cdot \cdots \cdot b$ as a product of "$n$ copies of $b$." It is easy to verify that this implies

$$b^m \cdot b^n = b^{m+n}.$$

One way to look at this is that the operation of addition among the positive integers $m$ and $n$ is "translated" into multiplication of $b^m$ and $b^n$. This seems like a nice property, and we would like to keep it when we extend the definition of raising to the power to other exponents. We would like the property $b^\alpha \cdot b^\beta = b^{\alpha+\beta}$ to remain valid for all real numbers $\alpha$ and $\beta$. If we could raise $b$ to the power 0 then we should have $b^n b^0 = b^{n+0} = b^n$, so our only choice is to define: $b^0 = 1$. This makes sense as the additive "identity" 0 is now translated to the multiplicative identity 1.

For negative exponents we want to have $b^{-n}b^n = b^{-n+n} = b^0 = 1$; thus we should define

$$b^{-n} = \frac{1}{b^n}.$$

This takes care of all of $\mathbb{Z}$, all integer exponents. How about $\mathbb{Q}$, all fractions of integers? For a fractional exponent $\frac{1}{2}$ we should have $b^{\frac{1}{2}}b^{\frac{1}{2}} = b^{\frac{1}{2}+\frac{1}{2}} = b^1 = b$, which means that $b^{\frac{1}{2}} = \sqrt{b}$. By similar arguments we can convince ourselves that the only reasonable way to define raising $b$ to a rational power is

$$b^{\frac{m}{n}} = \sqrt[n]{b^m} = \left(\sqrt[n]{b}\right)^m,$$

for any integers $m, n \in \mathbb{Z}$, with $n \geq 1$.

Let us consider a function $f(x) = b^x$. So far we have established how to evaluate this function for all rational values of $x$. Suppose that we plot on the graph all the points $(x, b^x)$, where $x \in \mathbb{Q}$ is rational. Then we "connect the dots." This is possible since the set $\mathbb{Q}$ of all rationals is dense on the real number line, that is, every real number can be approximated with any desired accuracy by the rationals. (This can be formalized by saying that the value $b^\alpha$ is a limit of the sequence of values $b^{\frac{m}{n}}$ for rational exponents, as the numbers $\frac{m}{n}$ approximate the irrational exponent $\alpha$ better and better.)

The graph of the function $f(x) = b^x$ lies above the $x$-axis and goes through the point $(0, 1)$ for every value of $b$. If $b > 1$, then the function is increasing rapidly (exponentially). To obtain the graph of $f(x) = b^x$ for $b < 1$, notice that $b^x = \left(\frac{1}{b}\right)^{-x}$. All we need to do is flip the graph of $g(x) = \left(\frac{1}{b}\right)^x$ about the $y$-axis.

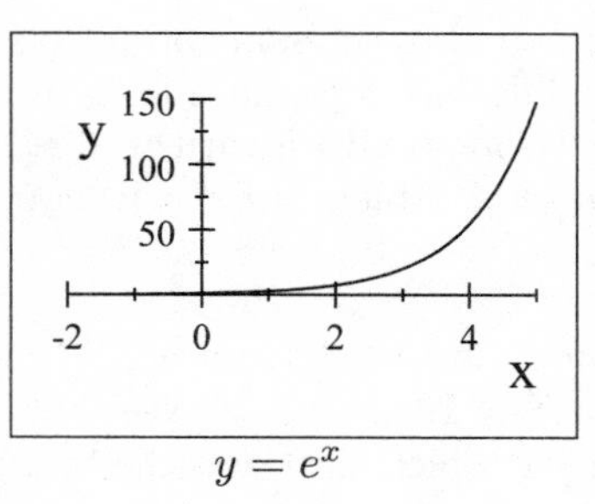


$y = e^x$

There is one value of the base $b$ which is of particular importance: $e \approx 2.718\,3$. The symbol for the constant $e$ honors Leonard Euler, who studied its properties extensively in the early eighteenth century. We will encounter this mysterious number many times in this book, but for some motivation, please feel free to look at Problems 10 and 11.

The logarithm of $x$ to the base $b$ is the exponent to which $b$ needs to be raised to get $x$, that is,

$$\log_b x = y \text{ if and only if } b^y = x.$$

In other words, $g(x) = \log_b x$ is the inverse function to the exponential function $f(x) = b^x$. For example, $10^3 = 1000$, which means that $\log_{10}(1000) = 3$. (Just in case the reader forgot, the inverse function has nothing to do with a reciprocal. It is the function that brings your $x$ back. In mathematics we can even do that for you!)

For $b > 1$, the function $f(x) = \log_b x$ is increasing (very slowly, at least for large $x$). Its graph always passes through the point $(1, 0)$, regardless of the value of the base $b$. It is commonly accepted, at least in the sciences, to use the notation

$$\begin{aligned} \log x &= \log_{10} x \quad \text{(decimal logarithm)}, \\ \ln x &= \log_e x \quad \text{(natural logarithm)}. \end{aligned}$$

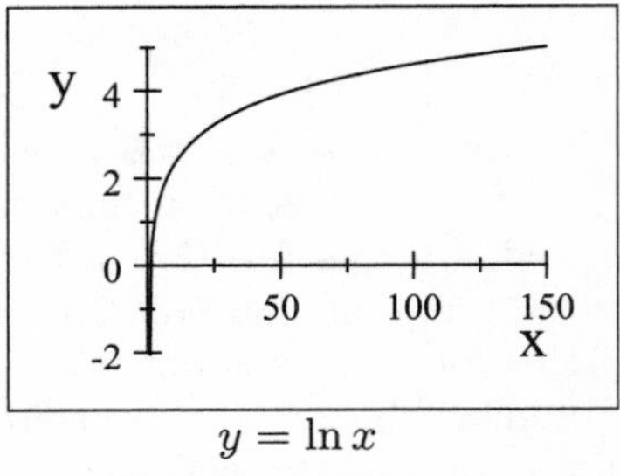


$y = \ln x$

Logarithms have important properties, which are equivalent to the corresponding properties of the exponentials. We list them next to each other:

$$\begin{array}{rcl} \log_b(xy) = \log_b x + \log_b y & \leftrightarrows & b^\alpha b^\beta = b^{\alpha+\beta}, \\ \log_b\left(\frac{x}{y}\right) = \log_b x - \log_b y & \leftrightarrows & \frac{b^\alpha}{b^\beta} = b^{\alpha-\beta}, \\ \log_b(x^r) = r\log_b x & \leftrightarrows & (b^\alpha)^\beta = b^{\alpha\beta}, \\ \log_b x = \frac{\log_c x}{\log_c b} = \frac{\log x}{\log b} = \frac{\ln x}{\ln b} & \leftrightarrows & (b^\alpha)^\beta = b^{\alpha\beta}, \end{array} \tag{1.8}$$

where $x, y > 0$, of course. Two of the formulas correspond to the same property of exponents and in fact follow from each other. The last formula allows us to rewrite a logarithm to a different base. When working with logarithms, it is crucial to use their properties correctly. Please see Problem 9 for a list of some common mistakes.

**Example 3** Let us calculate a few values of logarithms: $\log 10 = 1$, since $10^1 = 10$; $\log 100 = 2$, since $10^2 = 100$; $\log 1000 = 3$, as $10^3 = 1000$. We can see that $3 = \log 1000 = \log(100 \cdot 10) = \log 100 + \log 10 = 2 + 1$, which illustrates the first property in (1.8). As further examples, we can notice that $\ln e = 1$; $\log_b 1 = 0$, regardless of the base $b$; $\log_b(b^x) = x$; $\ln(e^x) = x = e^{\ln x}$. On the other hand $e^{2\ln x} = \left(e^{\ln x}\right)^2 = x^2$ (and it is not $2x$).

**Exercise 4** *Convince yourself that the formulas (1.8) are correct. For instance, suppose that* $\log_b x = \alpha$ *and* $\log_b y = \beta$. *Check that* $b^{\alpha+\beta} = xy$. *Rewrite it back into logarithms. For the change of base formula, start with* $y = \log_b x$, *which means* $x = b^y$. *Then take* $\log_c$ *of both sides.*

## 1.7 Trigonometric functions

Trigonometry plays an important role in calculus, not just because we may be interested in some geometric considerations. Trigonometric functions are useful in integration techniques, (Section 6.4). Also, the inverse trigonometric functions are antiderivatives of fairly simple functions, which we would otherwise not be able to integrate, (Section 3.4). Please consult [4] or [6] for a concise, but more complete treatment.

The values of $\cos\alpha$ and $\sin\alpha$, for any real number $\alpha$, are defined as the coordinates of the point on the unit circle at distance $\alpha$ units from $(1,0)$ measured counterclockwise along the arc. This way of looking at the sin and cos functions immediately explains the sign of each function in every quadrant. It also explains several reduction identities.

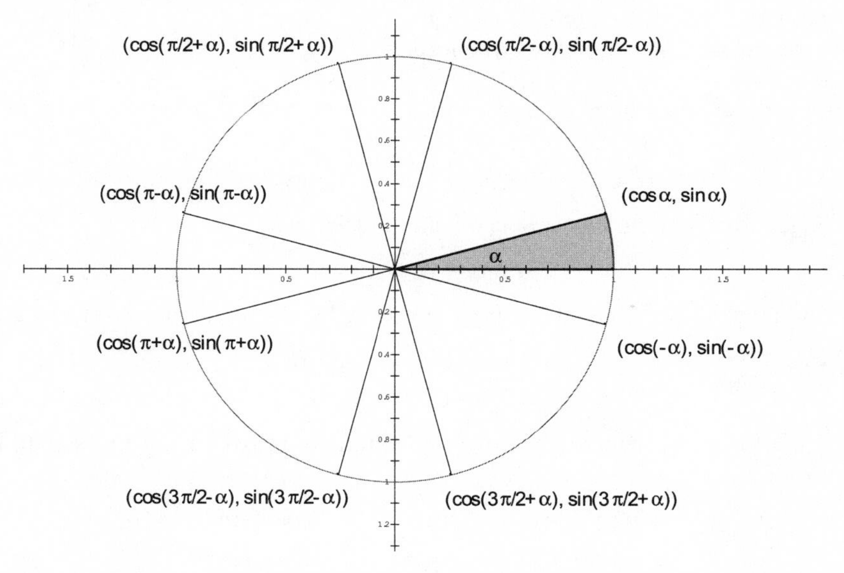


For example, to obtain the reduction formula $\sin\left(\alpha+\frac{\pi}{2}\right)=\cos\alpha$, all we need to do is compare the coordinates of appropriate points on the unit circle. Similarly, $\sin(-\alpha)=-\sin\alpha$ and $\cos(-\alpha)=\cos\alpha$, which means that sine is an odd function while cosine is even. The reduction formulas are useful, but there is no need to memorize them, as it takes just a few moments to sketch a simple picture. We assume that the reader is familiar with the graphs of trigonometric functions.

The values of the trigonometric functions for certain angles are frequently used and are good to know. The the values for $\alpha = 0$, $\pi$ or $\frac{\pi}{2}$ follow immediately from the definition: just visualize where on the unit circle the point $(\cos\alpha, \sin\alpha)$ is located. To obtain the values for $\alpha = \frac{\pi}{4}$, is it useful to cut a square in half by the diagonal. The resulting angles are $\frac{\pi}{4}$, $\frac{\pi}{4}$ and $\frac{\pi}{2}$. To get the values for $\frac{\pi}{3}$ and for $\frac{\pi}{6}$, start with an equilateral triangle and slice it in half.

There are many trigonometric formulas, but some of them occur quite frequently in calculus and are worth remembering. We list them here, along with ways to derive them quickly, whenever feasible.

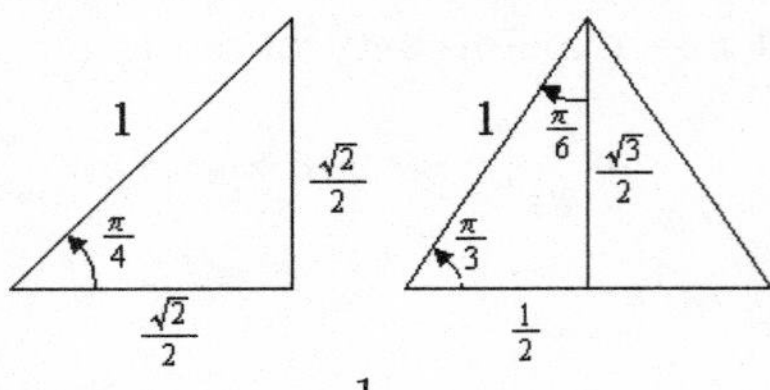


Definitions of other trigonometric functions:

$$\tan\alpha = \frac{\sin\alpha}{\cos\alpha}; \quad \sec\alpha = \frac{1}{\cos\alpha}; \quad \csc\alpha = \frac{1}{\sin\alpha}.$$

Pythagorean theorem for a point on the unit circle:

$$\sin^2\alpha + \cos^2\alpha = 1.$$

By dividing the previous identity by $\sin^2\alpha$ or by $\cos^2\alpha$, respectively, we get:

$$\begin{aligned} 1 + \cot^2\alpha &= \csc^2\alpha, \\ \tan^2\alpha + 1 &= \sec^2\alpha. \end{aligned}$$

Addition formulas, which are too complicated to derive on the fly (see [6] for the proof):

$$\begin{aligned} \sin(\alpha + \beta) &= \sin\alpha\cos\beta + \cos\alpha\sin\beta, \\ \cos(\alpha + \beta) &= \cos\alpha\cos\beta - \sin\alpha\sin\beta. \end{aligned}$$

To obtain the subtraction formulas, just replace $\beta$ with $-\beta$:

$$\begin{aligned} \sin(\alpha - \beta) &= \sin\alpha\cos\beta - \cos\alpha\sin\beta, \\ \cos(\alpha - \beta) &= \cos\alpha\cos\beta + \sin\alpha\sin\beta. \end{aligned}$$

Taking $\alpha = \beta$ in the addition formulas gives the double angle formulas:

$$\begin{aligned} \sin(2\alpha) &= 2\sin\alpha\cos\alpha, \\ \cos(2\alpha) &= \cos^2\alpha - \sin^2\alpha = 2\cos^2\alpha - 1 = 1 - 2\sin^2\alpha \end{aligned} \tag{1.9}$$

To switch between different versions of the last formula we use the Pythagorean identity: one can always replace $\sin^2\alpha$ with $1-\cos^2\alpha$ and $\cos^2\alpha$ with $1-\sin^2\alpha$. The double angle formulas are especially useful in integration.

The inverse trigonometric functions give us the angles whose corresponding trigonometric functions have specified values. However, trigonometric functions not invertible: if all we know is that $\sin\alpha = \frac{1}{2}$, then $\alpha$ could be $\frac{\pi}{6} + 2k\pi$ or $\frac{5\pi}{6} + 2k\pi$, where $k$ is some arbitrary integer. For this reason, we restrict the domains of the trigonometric functions before finding their inverses. Alternatively, we restrict the range of values for the inverse functions. The domains of the inverse trigonometric functions are the same as ranges of the original functions. We define:

$$\begin{aligned} &\arcsin(x) \text{ is the angle } \alpha \in \left[-\frac{\pi}{2}, \frac{\pi}{2}\right], \text{ with } \sin\alpha = x, \\ &\arccos(x) \text{ is the angle } \alpha \in [0, \pi], \text{ with } \cos\alpha = x, \\ &\arctan(x) \text{ is the angle } \alpha \in \left(-\frac{\pi}{2}, \frac{\pi}{2}\right), \text{ with } \tan\alpha = x. \end{aligned} \tag{1.10}$$

**Warning:** $\sin^2\alpha = (\sin\alpha)^2$, but $\sin^{-1}\alpha \neq (\sin\alpha)^{-1} = \frac{1}{\sin\alpha}$!!! OK, so we are a little inconsistent here, but that's because at some point someone could not fit the word "arcsin" on a calculator button. Just to be clear: $\sin^{-1}\alpha = \arcsin\alpha$. The $-1$ in the "exponent" stands for the inverse function and not for the multiplicative inverse of the number $\sin\alpha$.

## 1.8 Problems

1. Complete the squares to find the center and radius of the circle given by the equation: $x^2 + 6x + y^2 - 4y = 3$.

2. Find an equation of the line passing through points $P(3,4)$ and $Q(5,0)$.

3. Let $f : \mathbb{R} \to \mathbb{R}$ denote some real-valued function on the set of real numbers. Write the formula for the slope of the line passing through points:

   (a) $P(1, f(1))$ and $Q(3, f(3))$,

   (b) $P(a, f(a))$ and $Q(b, f(b))$,

   (c) $P(a, f(a))$ and $Q(a+h, f(a+h))$ (See Section 2.2).

4. Match the functions with their graphs (yes, four functions lose this game):

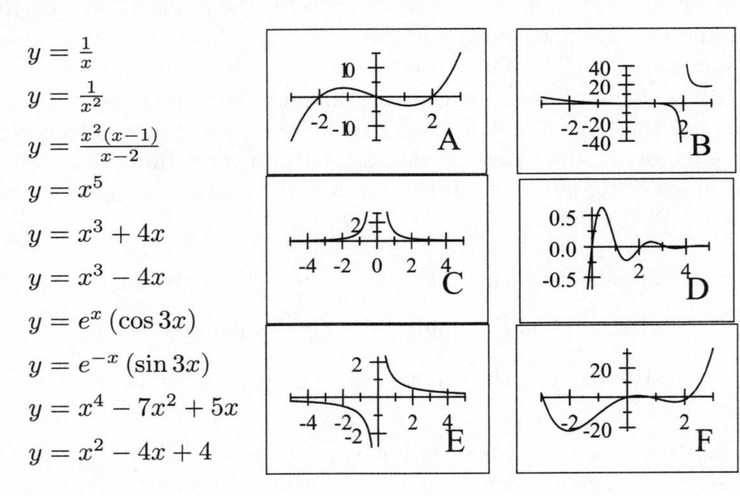


5. Find the domain of the function $f(x) = \sqrt{\frac{(x-3)^2(x-5)}{x^2(x-2)}}$.

6. Solve the inequality $|x-2| + |x-5| < 9$.

7. Find all solutions of the equation $x^3 - 2x + 1$. Hint: notice that $x = 1$ is one of them.

8. Convince yourself that for any positive numbers $x, b > 0$, we have $b^{\log_b x} = \log_b(b^x) = x$.

9. Check that the following are *really bad* ideas:

(a) $\log\left(\frac{x}{y}\right) = \frac{\log x}{\log y}$;

(b) $\log\sqrt{x} = \sqrt{\log x}$;

(c) $\log(xy) = (\log x)(\log y)$;

(d) $\sin(2x) = 2\sin x$;

(e) $\sin^2(3x) + \cos^2(3x) = 3$.

10. Graph the function $f(x) = (1+x)^{\frac{1}{x}}$ on a calculator, if you have one handy. Otherwise, evaluate it for a few values of $x$, close to 0. Observe that the values appear to approach $e$ as $x$ gets close to 0.

11. Consider an investment of one dollar (or one million dollars if it sounds better) in a bank which offers the annual interest rate of $r$. Suppose that the interest is compounded $m$ times per year.

    (a) What is the value of your investment after $t$ years if $m = 1$? How about $m = 12$ or 365?

    (b) Consider what happens if the interest is compounded continuously, that is, as $m \to \infty$. Hint: $\left(1+\frac{r}{m}\right)^{mt} = \left(1+\frac{r}{m}\right)^{\frac{m}{r}rt} = \left[\left(1+\frac{r}{m}\right)^{\frac{m}{r}}\right]^{rt}$. When $m \to \infty$, then $\frac{r}{m} \to 0$. Now look at the previous problem.

12. Find the value of $k$, such that $f(x) = 10^x = e^{kx}$. This problem illustrates that the graph of $y = 10^x$ can be obtained from the graph of $y = e^x$ by horizontal shrinking by $k$.

13. Show that $\sin\left(\frac{\pi}{2} - x\right) = \cos x$ in two ways: by using the trigonometric circle and by using the subtraction formula for $\sin(s-t)$.

# Chapter 2

# Limits and Derivatives, the Concept

## 2.1 Slope of a curve - rate of change

One of the central ideas in calculus is that a typical function locally looks like a straight line. In other words, if we "zoom in" on the graph of a "decent" function sufficiently close, it becomes indistinguishable from a straight line. This allows us to generalize the concept of a slope of a straight line to the **slope of a curve**. Specifically, the slope of a curve $y = f(x)$ at $x = x_0$ is the slope of the line tangent to the graph of $f$ at the point $(x_0, f(x_0))$. In other words, the **derivative** $f'(a)$ is the instantaneous rate of change of the function $f$ at a given value of $x = a$. We are taking for granted that we know what the **tangent line** is. We will redefine the derivative, without using the tangent line in Section 2.2.

A function is called **differentiable** (at a point $x$) if it has a derivative (at the point $x$). Such functions are locally well approximated by tangent lines. The simplest and most common reason for which a continuous function may not be differentiable is that its graph has a "cusp." For example, $f(x) = |x|$ is not differentiable at $x = 0$ (sketch a graph). A typical function in calculus is differentiable, except possibly at a few points.

What is "typical" is another story. Just as most numbers on the real line are not rational, most functions are not even continuous, not to mention differentiable. In fact, a randomly selected function is not likely to be continuous even at a single point. Moreover, among continuous functions, most are not differentiable anywhere, (an elegant proof of this fact uses the Baire Category theorem, but this is a whole other story). Luckily, most functions which we encounter in the sciences are much more smooth.

The positive (negative) value of the derivative $f'$ means that the function itself is increasing (decreasing), at least in the immediate vicinity of the point in question.

The value of the second derivative $f'' = (f')'$, that is, the derivative of the derivative, controls the rate of change of the first derivative $f'$. Suppose that $f''(x) > 0$ on some interval. Then the first derivative $f'$ (or slope of $f$) increases, which in turn implies that the graph turns upward. We say that the function $f$ is **concave up**. The graph of such a function lies locally above the tangent line.

Conversely, if $f''(x) < 0$ then the function $f$ is **concave down** (the graph turns to the right, or downward; the graph of $f$ is locally below the tangent line). This is summarized below:

| $f'$ | $-$ | $+$ | $+$ | $-$ |
|---|---|---|---|---|
| $f''$ | $+$ | $+$ | $-$ | $-$ |
| $f$ |  |  |  |  |

(2.1)

A number $x_0$ in the domain of $f$ is called:

- a **stationary** point of $f$ if $f'(x_0) = 0$ (some authors call these critical points);
- a **critical** point of $f$ if it is stationary or if $f'(x_0)$ does not exist;
- a **local (global) maximum** point of $f$ is $f(x) \leq f(x_0)$ in the immediate vicinity of $x_0$ (entire domain of $f$);
- a **local minimum (global)** point of $f$ if $f(x) \geq f(x_0)$ in the immediate vicinity of $x_0$ (entire domain of $f$);

- an **inflection** point if the concavity of $f$ changes at $x_0$.

It follows that if the graph of a function $f$ first goes up and then goes down, then $f$ has a local maximum, and vice versa. In other words, if the derivative $f'$ changes sign from negative to positive at $x_0$, then $f$ has a minimum at $x_0$, and the other way around. This is usually called:

**First derivative test.** Suppose that $f^{'}(x_0) = 0$. Let $\varepsilon$ be some (small) positive number.

- If $f'(x) < 0$ for $x_0 - \varepsilon < x < x_0$ and $f'(x) > 0$ for $x_0 < x < x_0 + \varepsilon$, then $f$ has a local minimum at $x_0$;
- If $f'(x) > 0$ for $x_0 - \varepsilon < x < x_0$ and $f'(x) < 0$ for $x_0 < x < x_0 + \varepsilon$, then $f$ has a local maximum at $x_0$.

Sometimes it is more convenient to look at concavity: if the function looks like a valley, it has a local minimum at a stationary point; if it looks like a hill, then it has a local maximum.

**Second derivative test.** Suppose that $f'(x_0) = 0$.

- If $f''(x_0) > 0$, then $f$ has a local minimum at $x_0$;
- If $f''(x_0) < 0$, then $f$ has a local maximum at $x_0$;
- If $f''(x_0) = 0$, then anything can happen (try going back to the first derivative test).

It is evident that the information about the first and second derivatives may tell us something about the function itself. In practice, we may only have the numerical information and not an actual formula - that's life. However, when we do have the formula for $f(x)$, we should be able to differentiate $f$ symbolically. Then we can use algebraic, graphical or numerical methods to gain information about $f'$ and $f''$. We will develop efficient methods of finding derivatives symbolically in the next chapter, but first we have to define derivative more carefully.

## 2.2 Formal definition of derivative

In order to calculate the instantaneous rate of change of the function, we calculate the average rate of change over a small interval and then see what happens when the interval becomes smaller and smaller. Let us pick two points on the graph of the function $P(a, f(a))$ and $Q(x, f(x))$, not too far from each other.

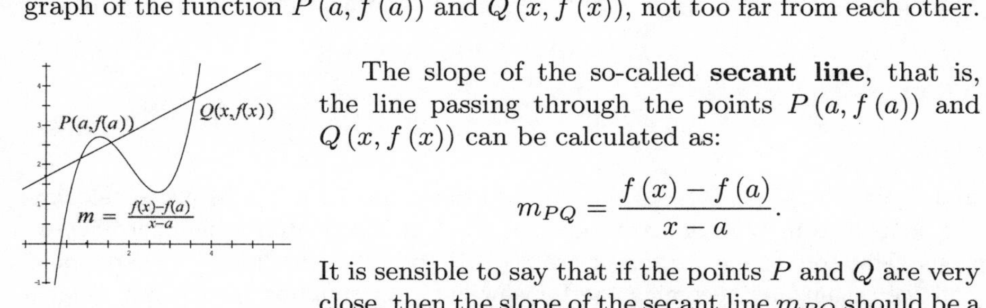


The slope of the so-called **secant line**, that is, the line passing through the points $P(a, f(a))$ and $Q(x, f(x))$ can be calculated as:

$$m_{PQ} = \frac{f(x) - f(a)}{x - a}.$$

It is sensible to say that if the points $P$ and $Q$ are very close, then the slope of the secant line $m_{PQ}$ should be a very good approximation of the slope of the *tangent line*. We call this slope the **derivative of the function** $f$ at $a$ and denote it $f'(a)$ (following Lagrange). A mathematical way of expressing this is

$$f'(a) = \lim_{x \to a} \frac{f(x) - f(a)}{x - a}. \tag{2.2}$$

Alternatively, if we denote $x = a + h$, then $h$ denotes how far it is from $a$ to $x$, (a "directed distance"). The derivative can then be defined as

$$f'(a) = \lim_{h \to 0} \frac{f(a+h) - f(a)}{h}. \tag{2.3}$$

We will need to define the concept of a limit precisely (coming up in Section 2.3), but for now let us look at a few examples, keeping in mind that the derivative $f'(a)$ is the slope of the line tangent to the graph of $f$ at the point $P(a, f(a))$. A point-slope form of the equation of the tangent line is

$$y - f(a) = f'(a) \cdot (x - a),$$

which is usually written as

$$y = f(a) + f'(a) \cdot (x - a). \tag{2.4}$$

**Example 5** Let $f(x) = x^2$ and $a = 2$. For any other point $Q(x, x^2)$ on the graph of $f$, we have

$$m_{PQ} = \frac{x^2 - 2^2}{x - 2} = \frac{(x-2)(x+2)}{x-2} = x + 2;$$

therefore $f'(2) = \lim_{x \to 2}(x+2) = 2 + 2 = 4$. Another approach is to write

$$\begin{aligned} m_{PQ} &= \frac{f(2+h) - f(2)}{h} = \frac{(2+h)^2 - 4}{h} \\ &= \frac{4 + 4h + h^2 - 4}{h} = \frac{h(4+h)}{h} = 4 + h, \end{aligned}$$

and again $f'(2) = \lim_{h \to 0}(4+h) = 4$. In this example, it is possible to calculate the limits by simply plugging the value of $x = 2$ or $h = 0$ into the simplified expressions. This corresponds to the points $P(2,4)$ and $Q(2+h, (2+h)^2)$ overlapping, and the secant line becoming the tangent line. An equation for the line tangent to the graph of $f$ at the point $(2,4)$ is $y = 4 + 4(x-2) = 4x - 4$.

It is clear from the example above that in order to calculate $f(a+h)$ correctly, we need to replace the variable in the formula for $f(x)$ with $a+h$. One of the more common mistakes beginners make is to replace $f(a+h)$ with $f(a) + h$ or with $f(a) + f(h)$.

**Exercise 6** *Repeat the argument above for a point $P(a, a^2)$ on the graph of the function $f(x) = x^2$ to obtain its derivative $f'(a) = 2a$.*

**Exercise 7** *Consider $f(x) = x^3$. Use the definition of the derivative to calculate the derivative at $f'(a)$ at $x = a$.*

**Exercise 8** *Try finding derivatives of other functions, such as $\sqrt{x}$, $\frac{1}{x}$ or $\frac{1}{x^2}$ (see Section 3.1 for solutions).*

Another commonly used notation for derivatives was introduced by Leibnitz. This notation takes more space, but is also more informative. We write:

$$\begin{aligned} f'(x) &= \frac{d}{dx}(f(x)) = \frac{df}{dx} \qquad \text{and } f'(a) = \left.\frac{df}{dx}\right|_{x=a}, \\ f''(x) &= \frac{d^2 f}{dx^2} = \frac{d^2}{dx^2}(f(x)) \text{ and so on.} \end{aligned}$$

Both notations are used interchangeably in this book.

## 2.3 Limits, continuity and differentiability

Our immediate motivation to develop the precise concept of the limit is that we would like to make the definition of the derivative precise. However, limits are one of the central concepts of calculus and are useful for other purposes as well, such as the proper definition of continuity or asymptotes. Instead of considering just limits of the form $\lim_{x\to a} \frac{f(x)-f(a)}{x-a}$ or $\lim_{h\to 0} \frac{f(a+h)-f(a)}{h}$, we will consider a general situation and define the limit of the function at a point: $\lim_{x\to a} f(x)$.

The general idea of a limit is that while we may not be able simply to plug in $x = a$ into the expression for $f(x)$ (or plugging it in may not give us the value we are interested in), we try to find a value $L$ to which the values of $f(x)$ are getting close when $x$ gets close to $a$, without actually becoming $a$. In other words, the limit of a function $f$ as $x$ approaches $a$, denoted as $\lim_{x\to a} f(x)$, is the value which the function "should have" at $a$. Whether the function is undefined at $x = a$, or is defined and has the "wrong" value, is irrelevant. Still somewhat informally we have:

**Definition 9** *We say that* $\lim_{x\to a} f(x) = L$ *if the values of* $f(x)$ *are close to* $L$*, when* $x$ *is close to* $a$*, but not equal to* $a$*.*

**Example 10** Let

$$g(x) = \begin{cases} 2x & \text{if } x \neq 1, \\ 5 & \text{if } x = 1. \end{cases}$$

The graph of $g(x)$ would be a straight line, except for the fact that someone took the point $(1, 2)$ from this line and moved it up to $(1, 5)$. The value of the function at $x = 1$ "should be" 2, which we express as $\lim_{x\to 1} g(x) = 2$.

**Example 11** Consider a function $f(x) = \frac{\sin x}{x}$.

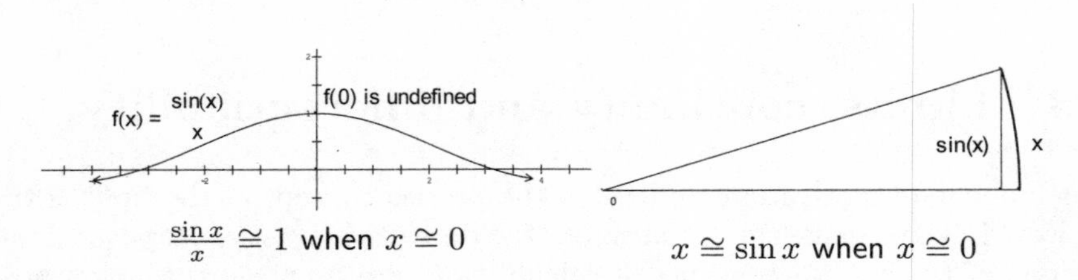


$\frac{\sin x}{x} \cong 1$ when $x \cong 0$ $\qquad$ $x \cong \sin x$ when $x \cong 0$

This time the value of the function has not been deliberately "spoiled" at $x = 0$, but it is not defined since cannot divide by 0. However, it seems clear from the graph of the function that $f(x)$ is close to 1, when $x$ is sufficiently close to 0. This makes sense, as the value of $\sin x$ is very close to the value of $x$, for small values of $x$. Therefore,

$$\lim_{x \to 0} \frac{\sin x}{x} = 1. \tag{2.5}$$

**Example 12** This time let $h(x) = \sin\left(\frac{1}{x}\right)$. It is difficult to get a decent looking graph of this function on a calculator or on a computer screen. It does not help to "zoom in," as we only get more and more oscillations on the screen.

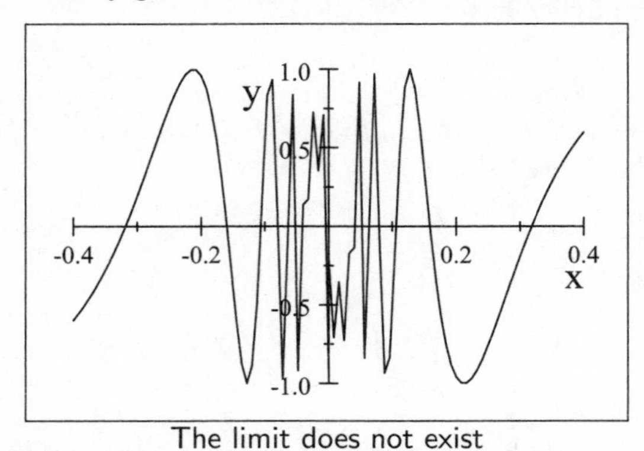


The limit does not exist

For this function the $\lim_{x \to 0} k(x)$ *does not exist.* This is because there is no single value $L$, such that $h(x)$ is close $L$, whenever $x$ is sufficiently close to zero. In fact, if $x_n = \frac{1}{n\pi}$, where $n = 1, 2, 3, \ldots$ is a positive integer, then $k(x_n) = \sin(n\pi) = 0$. On the other hand, if $t_n = \frac{1}{(n+1/2)\pi}$, where $n = 1, 2, \ldots$, then $k(t_n) = \sin\left(n\pi + \frac{\pi}{2}\right) = 1$.

The informal definition 9 of a limit relies on our intuitive understanding of what it means to "be close." The distance between any two numbers $a$ and $b$ can be expressed as $|a - b|$. To say that the number $f(x)$ is close to the number $L$ is to say that the distance from $f(x)$ to $L$ is smaller than $\varepsilon$, where $\varepsilon$ is some small number, that is, $|f(x) - L| < \varepsilon$. In the same way, to state that $x$ is close to $a$ but not equal to $a$, we can write $0 < |x - a| < \delta$, where $\delta$ is another small number. For the limit $\lim_{x\to a} f(x)$ to be $L$, we need to know that the values of $f(x)$ are within $\varepsilon$ from $L$, as long as $x$ is sufficiently close to $a$. This language leads to the formal definition of a limit.

**Definition 13** *We say that*

$$\lim_{x\to a} f(x) = L$$

*if for every* $\varepsilon > 0$, *there exists a* $\delta > 0$, *such that* $|f(x) - L| < \varepsilon$ *whenever* $0 < |x - a| < \delta$.

Let us notice that in definition 13 it is required that one can find a suitable $\delta$ for every possible choice of $\varepsilon$. The choice of $\delta$ is likely to depend on the initial choice of $\varepsilon$. The point is that it is always possible to find the allowed "tolerance" value $\delta$ for the input $x$, such that $f(x)$ must land in the target interval $(L - \varepsilon, L + \varepsilon)$.

The definition of the limit of a function took several years to be developed to its current form. The student should not feel intimidated by the fact that the definition 13 would not perhaps be his or her first thought. The $\varepsilon$-$\delta$ formalism is necessary when one needs to prove theorems rigorously. While in a typical calculus course we do not usually require such rigor, it is good to be aware of its necessity if one wants to make formal arguments about limits.

Having formally defined the concept of a limit, we now have the language to properly define continuity of a function. Informally, continuity at a point $x = a$ means that the graph of $f$ is not "broken" at $a$. We say that $f$ is **continuous** at $x = a$, if $\lim_{x\to a} f(x) = f(a)$. We say that $f$ is continuous on an interval $I$

if it is continuous at every $a \in I$. Informally, this means that one can sketch a graph of the function *without lifting a pencil.*

While continuity of $f$ at $x = a$ is an innocent looking statement, there really are three facts included in it:

1. $f(a)$ is defined, that is, $a$ must be in the domain of $f$.

2. $\lim_{x \to a} f(x)$ exists.

3. The two numbers: $\lim_{x \to a} f(x)$ and $f(a)$ are the same.

**Exercise 14** *Draw sample graphs of functions which illustrate how each of the conditions stated above can fail.*

Continuity of a function is a weaker condition than differentiability (having derivative). This should not come as a surprise: how can we find a line tangent to the graph if the graph is "broken." More formally speaking, suppose that a function $f$ is differentiable, that is, that the limit of a difference quotient (2.2) exists. The denominator $x - a$ is close to zero for $x$ close to $a$. In order for the limit of the ratio $\frac{f(x)-f(a)}{x-a}$ to have any chance to exist, the numerator $f(x) - f(a)$ must approach zero when $x$ is close to $a$. That means that $f$ is continuous at $a$.

Continuous functions have a property known as the **Intermediate Value Theorem**, which is easy to state and to believe, but much harder to prove. A graph of a continuous function such that $f(0) = -3$ and $f(1) = 5$ must have an $x$-intercept somewhere between $x = 0$ and $x = 1$. In general, if $f$ is continuous on a closed interval $[a, b]$, then it must hit all the values between $f(a)$ and $f(b)$. Formally, for every $y$ between $f(a)$ and $f(b)$, there exists at least one value of $c \in [a, b]$, such that $f(c) = y$.

Differentiable functions satisfy another "existence" theorem. Suppose that $f$ is differentiable on $[a, b]$. The average rate of change of $f$ over the entire interval is the slope of the secant line: $m_{[a,b]} = \frac{f(b)-f(a)}{b-a}$. There exists at least one value of $c \in (a, b)$ such that $f'(c) = m_{[a,b]}$. This is known as the **Mean Value Theorem.** Again, the proof is beyond the level of this book, but if it takes us 2 hours to travel 130 miles, it is reasonable to believe that our speed was exactly 65 m.p.h. at some point during the trip.

Another term which we can now properly define is the **tangent line**. Until now, we have been using an informal notion of the tangent line and we understood the derivative as the slope of that line. This process can now be reversed: by definition, we will say that the line tangent to the graph of $f$ at the point $(a, f(a))$ is the line passing through this point which has the slope $m = f'(a)$, as defined by equation (2.2) or (2.3). Therefore, the tangent line, by definition, is the line given by the equation (2.4).

Again, this formalization of seemingly simple ideas of continuity and differentiability took a long time to be developed. In the early stages of calculus, it was usually just assumed that all functions are continuous and differentiable. But then, all kinds of examples of functions were found, showing that the matter is more delicate. Charles Hermite was reported to say things like "I turn away with fright and horror from the evil of functions with no derivatives."

We end this section with a word of caution: just because one cannot plug $a$ into $f$ does not mean that the limit does not exist. This is why limits were invented in the first place! Yes, for continuous functions we have $\lim_{x\to a} f(x) = f(a)$, but in general plugging in the value of $a$ for $x$ does not work. If plugging $x = a$ gives, for example, $\frac{0}{0}$, that means not that the limit doesn't exist, only that there is more work to be done. Anything can happen. For instance, consider three limits of this type: $\lim_{x\to 0} \frac{3x}{x} = 3$, $\lim_{x\to 0} \frac{3x^2}{x} = 0$ and $\lim_{x\to 0} \frac{|x|}{x}$ does not exist (the values approach 1 or $-1$, depending on the sign of $x$).

When we calculate the derivative from the definition, we routinely run into expressions which would look like $\frac{0}{0}$ if we simply plugged in $h = 0$. But we do not plug it in! It is important not to jump to conclusions, when we encounter an indeterminate expression. We will come back to this matter in Section 4.1.

## 2.4 Problems

1. Sketch a graph of a function $f$, defined for all real numbers, such that $0 < f(x) \leq 1$ for all $x$, $f'(x) > 0$ for $x < 0$, $f'(x) < 0$ for $x > 0$, $f''(x) < 0$ for $x \in (-1, 1)$ and $f''(x) > 0$ for all other values of $x$.

2. Suppose that some function $g$ has derivative $g'(x) = \frac{x^2(x+3)}{x-5}$. Find all the stationary points for this function. Does $g$ have a local minimum

(maximum)? If so, at what values(s) of $x$? What happens at $x = 0$?

3. A graph of the derivative $p'(x)$ of a function $p$, whose domain is $[0, 5]$ is given.

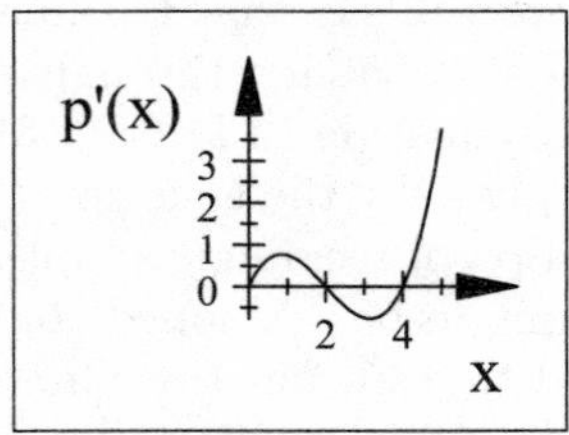


   (a) Suppose that $p(1) = 3$. Find the equation of the line tangent to the graph of $p$ at the point $(1, 3)$.
   (b) Use this information to approximate the value of $p(1.1)$.
   (c) On what interval is the function $p$ increasing (decreasing)?
   (d) Does $p$ have any stationary points? If so, does it have a maximum (minimum) point?
   (e) Approximately on what interval(s) is the function $p$ concave up?
   (f) Sketch a possible graph of $p(x)$ and of $p''(x)$.

4. Repeat the previous exercise if $p'(x) = (x+1)^2 (x-2)$ and the domain is $[-2, 4]$. A graphing device can be used for part (e).

5*. Determine whether a function

$$f(x) = \begin{cases} x \sin\left(\frac{1}{x}\right) & \text{for } \quad x \neq 0, \\ 0 & \text{for } \quad x = 0 \end{cases}$$

is continuous. Without trying to be too formal, explain your answer to yourself or to a friend. (Please do not lose friends over this.)

6*. (For enthusiasts) Determine the set of points at which the function is continuos:

$$g(x) = \begin{cases} \frac{1}{q} & \text{if} \quad x = \frac{p}{q},\ p \geq 0, \text{ and the fraction} \\ & \qquad \text{is in the lowest integer terms,} \\ 0 & \text{if} \quad x \text{ is irrational.} \end{cases}$$

# Chapter 3

# How to Find Derivatives

## 3.1 Derivatives of some common functions

### 3.1.1 Power functions and polynomials

The derivative of a function is supposed to be a generalization of the concept of the slope of a straight line. That is, if $f(x) = mx + b$ then we should have $f'(x) = m$ for all $x$. Indeed, if we use equation (2.3), then $f'(x) = \lim_{h\to 0} \frac{m(x+h)+b-(mx+b)}{h} = \lim_{h\to 0} \frac{mh}{h} = m$.

We have seen in Section 2.2 that $\frac{d}{dx}\left(x^2\right) = 2x$. We also have:

$$\begin{aligned}
\frac{d}{dx}\left(x^3\right) &= \lim_{h\to 0} \frac{f(x+h)-f(x)}{h} = \lim_{h\to 0} \frac{(x+h)^3 - x^3}{h} \\
&= \lim_{h\to 0} \frac{x^3 + 3x^2h + 3xh^2 + h^3 - x^3}{h} \\
&= \lim_{h\to 0} \frac{h\left(3x^2 + 3xh + h^2\right)}{h} \\
&= \lim_{h\to 0} 3x^2 + 3xh + h^2 = 3x^2,
\end{aligned}$$

$$\begin{aligned}
\frac{d}{dx}\left(\frac{1}{x}\right) &= \lim_{h\to 0}\frac{\frac{1}{x+h}-\frac{1}{x}}{h} = \lim_{h\to 0}\frac{x-(x+h)}{hx(x+h)} = \lim_{h\to 0}\frac{-h}{hx(x+h)} \\
&= \lim_{h\to 0}\frac{-1}{x(x+h)} = -\frac{1}{x^2} = -x^{-2},
\end{aligned}$$

$$\begin{aligned}
\frac{d}{dx}\left(\sqrt{x}\right) &= \lim_{h\to 0}\frac{\sqrt{x+h}-\sqrt{x}}{h} \\
&= \lim_{h\to 0}\frac{\sqrt{x+h}-\sqrt{x}}{h}\cdot\frac{\sqrt{x+h}+\sqrt{x}}{\sqrt{x+h}+\sqrt{x}} \\
&= \lim_{h\to 0}\frac{x+h-x}{h\left(\sqrt{x+h}+\sqrt{x}\right)} \\
&= \lim_{h\to 0}\frac{1}{\sqrt{x+h}+\sqrt{x}} = \frac{1}{2\sqrt{x}} = \frac{1}{2}x^{-2}.
\end{aligned}$$

In general

$$\frac{d}{dx}\left(x^{\alpha}\right) = \alpha x^{\alpha-1}, \tag{3.1}$$

which is known as the **power rule**. We have gathered some circumstantial evidence for it for now. A short and elegant proof is included in Example 25 in Section 3.5.

**Exercise 15** *Show from the definition of derivative that for any differentiable functions $f$ and $g$, and for any constant $c$, we have*

$$\begin{aligned}
(f+g)' &= f' + g', \\
(c\cdot f)' &= c\cdot f'.
\end{aligned} \tag{3.2}$$

While the formalities of the proof are left to the reader, let us notice that the linearity properties (3.2) are easy to see from the graphical point of view: multiplying a function by a constant $c$ stretches the graph vertically by a factor of $c$. As the result, the slope at every value of $x$ is increased by the same factor. Similarly, to obtain the graph of $f+g$, one must add the values of $f(x)$ and $g(x)$ at every point. The rate of change of $f+g$ should be the sum of the rates of change of the two functions.

Any operation which has properties (3.2) is called **linear**. Now, by using linearity of differentiation we can take derivatives of polynomial functions: for instance, $\frac{d}{dx}\left(3x^4 - 5x^2\right) = 12x^3 - 10x$.

### 3.1.2 Exponential and logarithmic functions

By numerical experimentation with the slopes of secant lines passing through the points $(0,1)$ and $\left(h, 3^h\right)$, it is easy to convince ourselves that that the derivative of $g(x) = 3^x$ at $x = 0$ is approximately 1.1. Similarly, for $k(x) = 2^x$ we have $k'(0) \cong 0.7$.

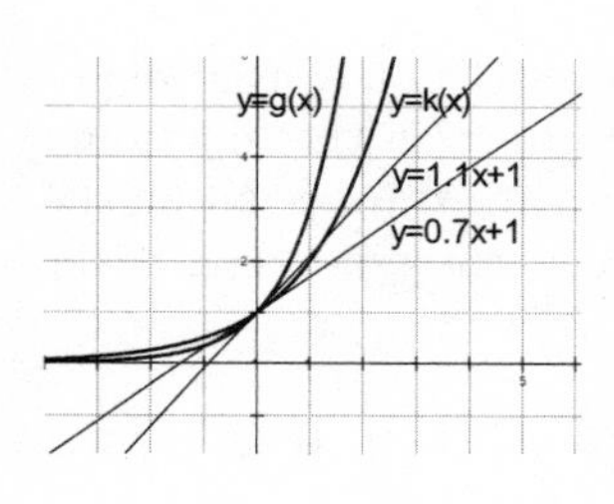


It is natural to expect that for some value of the base $b$, between 2 and 3, the derivative of the exponential function $b^x$ at $x = 0$ is exactly 1. This happens to be the same number $e \cong 2.7183...$ which we already encountered in Section (1.6) and in problems 10 and 11 after Chapter 1. This property of $e$ can be used as one of its several equivalent definitions: $e$ **is the value of the base** $b$ **of the exponential function** $f(x) = b^x$**, such that** $f'(0) = 1$.

To differentiate $f(x) = e^x$ we use the definition of derivative (see p. 19):

$$\begin{aligned}\frac{d}{dx}\left(e^x\right) &= \lim_{h\to 0}\frac{e^{x+h} - e^x}{h} = \lim_{h\to 0}\frac{e^x e^h - e^x}{h} = \lim_{h\to 0}\frac{e^x\left(e^h - 1\right)}{h} \\ &= e^x \lim_{h\to 0}\frac{e^h - e^0}{h} = e^x \cdot f'(0) = e^x \cdot 1 = e^x.\end{aligned}$$

For an arbitrary base $b$, we notice that $b^x = \left(e^{\ln b}\right)^x = e^{(\ln b)\cdot x}$. For any function, multiplying the variable $x$ by a constant $c$, in this case $c = \ln b$, shrinks the graph horizontally by the factor $c$ (see Appendix A.2). As a result, the slopes of the corresponding tangent lines get multiplied by $c$:

$$\left(b^x\right)' = b^x \ln b, \text{ and in particular } \left(e^x\right)' = e^x. \tag{3.3}$$

This is actually a special case of the chain rule (see Section 3.2.2). We will see yet another explanation of formula (3.3) in Section 3.3.

This property of the exponential function $e^x$ is the main reason it plays a fundamental role in all of mathematics. For example, once we have a function which is its own derivative, we can obtain solutions of differential equations describing radioactive decay, growth of bacteria (with an unlimited food supply) and accumulation of capital when the interest is compounded continuously. The common feature of all of these phenomena is that the rate of growth (or decay) is proportional to the current value of $y$, which can be described by a simple differential equation:

$$y' = k \cdot y. \tag{3.4}$$

Let $A$ and $k$ be constants. Then the exponential function $y = Ae^{kt}$ solves the initial value problem $y' = ky, \quad y(0) = A$ (see Section 7.4).

Let us now find the derivative of $y = \ln x$. We will use the fact that exponential and logarithmic functions are inverses of each other (in the functional sense, not as reciprocals, of course). Let $P(a, \ln a)$ be an arbitrary point on the graph of the logarithmic function $f(x) = \ln x$. There is corresponding point $Q(\ln a, a)$ on the graph of $g(x) = e^x$.

Due to the symmetry of the graphs, the slopes of the tangent lines at the points $P$ and $Q$ are reciprocals of each other. All we have to notice is that, at any point, the derivative of the exponential function matches the value of the $y$ coordinate of that point: $g'(x) = e^x = g(x)$. Specifically, at point $Q$, we have $g'(\ln a) = a$. Therefore, the slope of the line tangent to the graph of the logarithmic function at point $P$ has the slope $m = \frac{1}{a}$. We have

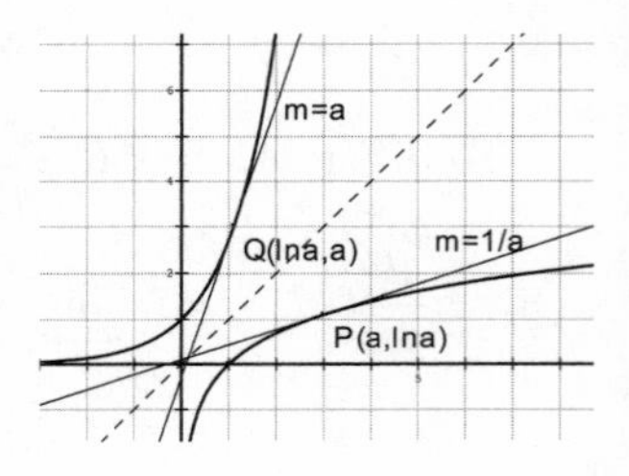


$$(\ln x)' = \frac{1}{x}, \text{ and in general } (\log_b x)' = \frac{1}{\ln b} \cdot \frac{1}{x}. \tag{3.5}$$

The second equality is easily obtained from the first, which is its special case for $b = e$, by using the change of base formula (see Section 1.6): $\log_b x = \frac{\ln x}{\ln b} = \frac{1}{\ln b} \cdot \ln x$ and the linearity of differentiation.

The logarithmic and exponential functions occur in various contexts throughout mathematics. It is a good idea to become comfortable with them instead of trying to avoid them.

### 3.1.3 Trigonometric functions

We will show that:

$$\begin{aligned} (\sin x)' &= \cos x, \\ (\cos x)' &= -\sin x. \end{aligned} \tag{3.6}$$

Let us start with the definition of derivative and apply the addition formula for sine:

$$\begin{aligned} (\sin x)' &= \lim_{h\to 0} \frac{\sin(x+h) - \sin x}{h} \\ &= \lim_{h\to 0} \frac{\sin x \cos h + \cos x \sin h - \sin x}{h} \\ &= \sin x \cdot \lim_{h\to 0} \frac{\cos h - 1}{h} + \cos x \cdot \lim_{h\to 0} \frac{\sin h}{h}. \end{aligned}$$

The expression $\lim_{h\to 0} \frac{\cos h - 1}{h}$ happens to be exactly the derivative of the function $y = \cos x$, at $x = 0$. By looking at the graph of cosine, it is easy to agree that $y'(0) = 0$. The second limit, $\lim_{h\to 0} \frac{\sin h}{h} = 1$, was already found in Section 2.3. We have

$$(\sin x)' = \sin x \cdot 0 + \cos x \cdot 1 = \cos x.$$

In order to find the derivative of cosine, we can rewrite it as $\cos x = \sin\left(x + \frac{\pi}{2}\right)$. Shifting a graph horizontally does not affect the slopes at corresponding points. Thus:

$$(\cos x)' = \left[\sin\left(x + \frac{\pi}{2}\right)\right]' = \cos\left(x + \frac{\pi}{2}\right) = -\sin x.$$

**Exercise 16** *Obtain the formula for the derivative of cosine from the definition, using the addition formula for* $\cos(x+h)$.

It is worth noting that formulas (3.6) are valid only if $x$ is in radians. If $x$ was in degrees, it would have to be converted to radians by multiplying it by $\frac{\pi}{180}$. For instance $[\sin(x^\circ)]' = \left[\sin(\frac{\pi}{180}x)\right]' = \frac{\pi}{180}\cos\left(\frac{\pi}{180}x\right) = \frac{\pi}{180}\cos(x^\circ)$. This is again a special case of the chain rule (Section 3.2.2), and it can be visualized by shrinking the graph of sine horizontally by the factor of $\frac{180}{\pi}$.

It is natural in this context to consider motion of a mass on a spring with no friction, called **simple harmonic motion**. If $y$ is a position of an object at the time $t$, then Newton's second law of motion implies: $y^{''} = F/m$, where $F$ is the force acting on the mass $m$. On the other hand, Hooke's law gives $F = -k_0 y$, which leads to a differential equation (see section 7.4) $y^{''} = -ky$, where $k = k_0/m$. For any constants $a$ and $b$,

$$y = a \sin\left(\sqrt{k}x\right) + b \cos\left(\sqrt{k}x\right) \tag{3.7}$$

is a solution to this differential equation. By choosing the right values of the coefficients $a$ and $b$, we can make sure that the initial conditions are satisfied (see problems 6 and 7 at the end of this chapter; also see Section 7.4).

We end this section by stating that

$$\begin{aligned} (\tan x)^{'} &= \sec^2 x, & (\sec x)^{'} &= \sec x \tan x, \\ (\cot x)^{'} &= -\csc^2 x, & (\csc x)^{'} &= -\csc x \cot x, \end{aligned} \tag{3.8}$$

all of which can be verified using the quotient rule from the next section.

## 3.2 Algebraic combinations of functions

### 3.2.1 Product and quotient rules

In order to see how to differentiate a product of two functions, let us consider a rectangle whose sides are two differentiable functions of time, say $f(t)$ and $g(t)$. The area of the rectangle $A(t) = f(t) \cdot g(t)$. When both sides of the rectangle are changing in time, the area $A(t)$ is changing "on two fronts." Let us assume that the left lower corner of the rectangle does not move. The rate with which the right edge of the rectangle sweeps the area is $f'(t) \cdot g(t)$. Similarly, the rate with which the area is swept by the upper edge of the rectangle is $f(t) \cdot g'(t)$. This leads to the **product rule**:

$$(fg)^{'} = f^{'}g + fg^{'}. \tag{3.9}$$

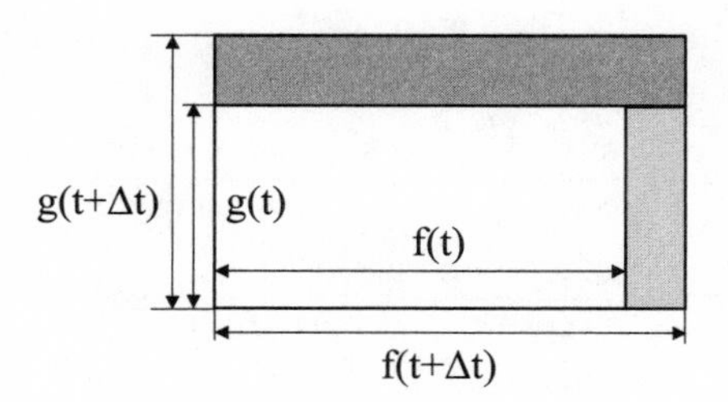


Product rule

Formally speaking, let us calculate the change of the area in a small increment of time $\Delta t$:

$$\Delta A = A\left(t+\Delta t\right) - A\left(t\right) = \Delta f\left(t\right)\cdot g\left(t+\Delta t\right) + f\left(t\right)\cdot \Delta g\left(t\right),$$

where $\Delta f\left(t\right) = f\left(t+\Delta t\right) - f\left(t\right)$ and $\Delta g\left(t\right) = g\left(t+\Delta t\right) - g\left(t\right)$. Dividing by $\Delta t$, we get

$$\begin{aligned}\frac{\Delta A}{\Delta t} &= \frac{\Delta f\left(t\right)\cdot g\left(t+\Delta t\right) + f\left(t\right)\cdot \Delta g\left(t\right)}{\Delta t} \\ &= \frac{\Delta f\left(t\right)}{\Delta t}\cdot g\left(t+\Delta t\right) + f\left(t\right)\cdot \frac{\Delta g\left(t\right)}{\Delta t},\end{aligned}$$

and taking the limit as $\Delta t \to \infty$ we obtain the product rule (3.9).

Having established the product rule by the above argument, we use it to verify the **quotient rule**:

$$\left(\frac{f}{g}\right)' = \frac{f'g - fg'}{g^2}. \tag{3.10}$$

Suppose that a function $u$ is a quotient $u = \frac{f}{g}$. Then $f = u\cdot g$, and by the product rule $f' = u'g + ug'$. Solving this equation for $u'$ we obtain

$$u' = \frac{f' - ug'}{g} = \frac{f' - \frac{f}{g}g'}{g} = \frac{f'g - fg'}{g^2}.$$

As an illustration of the quotient rule, let us verify some of the formulas (3.8):

$$\begin{aligned}(\tan x)' &= \left(\frac{\sin x}{\cos x}\right)' = \frac{\cos x \cdot \cos x - \sin x \cdot (-\sin x)}{\cos^2 x} \\ &= \frac{\cos^2 x + \sin^2 x}{\cos^2 x} = \frac{1}{\cos^2 x} = \sec^2 x.\end{aligned}$$

**Exercise 17** *Verify the remaining three formulas (3.8) using the quotient rule.*

**Exercise 18** *Use the "Freshman's product rule" $(f \cdot g)' = f' \cdot g'$ for $f(x) = x$ and $g(x) = x^2$. Compare the result with the derivative of $y = x^3$.*

**WARNING:** It is not possible to cancel $g$ in the quotient rule:

$$\frac{f'g - fg'}{g^2} \neq \frac{f' - fg'}{g}.$$

I apologize for stating the obvious, but this is a frequent mistake of calculus novices, who sometimes cannot resist the temptation to cross out the $g$ after differentiating a quotient. For example:

$$\left(\frac{x}{x^2+1}\right)' = \frac{1 \cdot (x^2+1) - x \cdot 2x}{(x^2+1)^2} \neq \frac{1 - 2x^2}{x^2+1}.$$

### 3.2.2 Composition and the chain rule

The **composition** of two functions $f$ and $g$ is defined as

$$(f \circ g)(x) = f(g(x)).$$

As is clear from the definition, composition has nothing to do with the product of the two functions. What happens is that the result from function $g$ is "fed" into the function $f$. Once again, there is no multiplication.

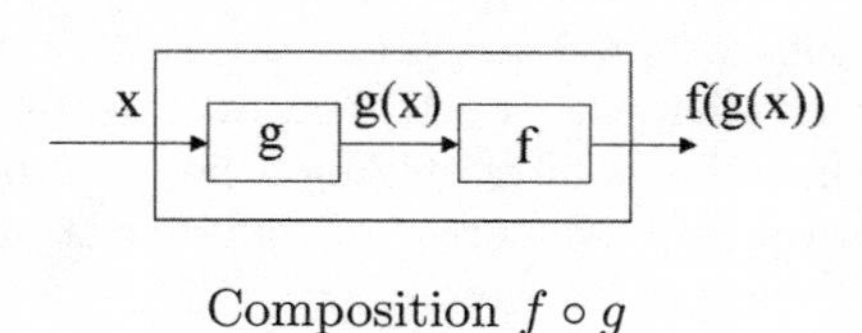


Composition $f \circ g$

If the functions composed are "nice," then their composition frequently has the same kind of "niceness." In particular, continuity is preserved, as is differentiability or even more restrictive types of smoothness.

In particular, the composition of differentiable functions is differentiable (at the appropriate points) and we have

$$\frac{d}{dx}(f \circ g)(x) = f'(g(x)) \cdot g'(x). \tag{3.11}$$

This rule is not unreasonable: if $f$ changes twice as fast as $g$ and $g$ changes three times as fast as $x$, then we would expect the composition to change six times as fast as $x$.

For those who are interested, here is a "proof":

$$\begin{aligned}
\frac{d}{dx}(f \circ g)(x) &= \lim_{h \to 0} \frac{f(g(x+h)) - f(g(x))}{h} \\
&= \lim_{h \to 0} \frac{f(g(x+h)) - f(g(x))}{g(x+h) - g(x)} \cdot \frac{g(x+h) - g(x)}{h} \\
&= \lim_{h \to 0} \frac{f(g(x+h)) - f(g(x))}{g(x+h) - g(x)} \cdot \lim_{h \to 0} \frac{g(x+h) - g(x)}{h} \\
&= f'(g(x)) \cdot g'(x).
\end{aligned}$$

This is more or less why the chain rule is true. Strictly speaking, we have no guarantee that the denominator $g(x+h) - g(x)$ is not zero! This technicality would have to be worked around in a rigorous proof.

Listed below are some special cases of the chain rule, which follow easily from the general version

$$\begin{aligned}
\left((g(x))^k\right)' &= k \cdot (g(x))^{k-1} \cdot g'(x), \\
\left(e^{g(x)}\right)' &= e^{g(x)} \cdot g'(x), \\
(\ln(g(x)))' &= \frac{g'(x)}{g(x)}.
\end{aligned} \tag{3.12}$$

Please note in particular that $\frac{d}{dx}\left(e^{g(x)}\right) \neq e^{g(x)}$ and $\frac{d}{dx}\left(\ln\left(g\left(x\right)\right)\right) \neq \frac{1}{g(x)}$. I ask the reader's forgiveness for making such an obvious comment. However, this type of omission often makes its way into students' work. For most people, learning to differentiate efficiently and correctly requires solving a large number of differentiation problems. There are plenty of sources of differentiation problems and this book contains just a few after this chapter.

**Example 19** Here is a quick list of examples, along with warnings for frequent errors:

$\frac{d}{dx}\left(e^{3x}\right) = e^{3x} \cdot 3 = 3e^{3x}$, but it is not just $e^{3x}$;

$\frac{d}{dx}\left(\ln\left(\cos x\right)\right) = \frac{1}{\cos x} \cdot \left(-\sin x\right) = -\tan x$, but not just $\frac{1}{\cos x}$;

$\frac{d}{dx}\left[\left(3x^2 - 7x + 5\right)^{10}\right] = 10\left(3x^2 - 7x + 5\right)^9\left(6x - 7\right)$, and not $10\left(6x - 7\right)^9$;

$\frac{d}{dx}\left(\frac{e^x}{x}\right) = \frac{e^x \cdot x - e^x \cdot 1}{x^2} = \frac{e^x(x-1)}{x^2}$;

$\frac{d}{dx}\left[\ln\left(x \sin x\right)\right] = \frac{1}{x \sin x}\left(1 \cdot \sin x + x \cdot \cos x\right) = \frac{\sin x + x \cos x}{x \sin x}$;

$\frac{d}{dx}\left[\tan\left(x^2 + 3x\right)\right] = \sec^2\left(x^2 + 3x\right) \cdot \left(2x + 3\right) = \left(2x + 3\right)\sec^2\left(x^2 + 3x\right)$.

## 3.3 Implicit functions and implicit differentiation

Suppose that $y$ is a function of $x$, but we do not have an explicit formula for $y$ in terms of $x$. All we have is an equation which $x$ and $y$ must satisfy. We would like to find an expression for the derivative $y'$. There is a price to pay for not solving the original equation for $y$: the expression for $y'$ will in general involve both $x$ and $y$. Even if not perfect, as it is not given explicitly in terms of a single variable, a formula like that can still be useful.

The key to success in implicit differentiation is to remember at all times that $y$ is a function of $x$, and not just a constant or another independent variable. What seems to lead to some confusion is an apparent inconsistency: $\left(x^2\right)' = 2x$ and yet $\left(y^2\right)' = 2y \cdot y'$. The prime notation, while concise, is not very informative. Using Leibnitz notation clarifies this issue: $\frac{d}{dx}\left(x^2\right) = 2x$

but $\frac{d}{dx}\left((y(x))^2\right) = 2y(x) \cdot \frac{dy}{dx}$. In practice, it is sufficient to remember, which letter is the variable and which is the function.

There is another interpretation of the variables, where $x$ and $y$ are both functions of time $t$. This will be addressed in Section 4.3 about the so-called related rates.

**Example 20** Assuming that $y$ is a function of $x$ given implicitly by the equation

$$x^3 + y^3 - x^2 \cdot y = 7,$$

find $y'$. First, we take the derivative of both sides, applying the chain rule to $y^3$ and the product rule to $x^2 \cdot y$ :

$$3x^2 + 3y^2 \cdot y' - \left(2x \cdot y + x^2 \cdot y'\right) = 0.$$

We put all the terms with $y'$ on one side of the equation and all leave the rest the other:

$$\begin{aligned} 3y^2 \cdot y' - x^2 \cdot y' &= 2x \cdot y - 3x^2, \\ y' \cdot \left(3y^2 - x^2\right) &= 2x \cdot y - 3x^2; \end{aligned}$$

after factoring $y'$, we divide both sides of the equation by the expression in parentheses to obtain

$$y' = \frac{2x \cdot y - 3x^2}{3y^2 - x^2}.$$

In practice, when we need to find a slope of the tangent line (that is, $y'$) only at a specific point, it may be easier to replace $x$ and $y$ by the appropriate values as soon we differentiate both sides of the equation.

**Example 21** Suppose that $y = \ln x$ and we have forgotten the formula for the derivative of $\ln x$. We have, from the definition of the logarithm,

$$e^y = x.$$

Differentiating both sides (with respect to $x$), we obtain

$$e^y \cdot y' = 1,$$

thus

$$y' = \frac{1}{e^y} = \frac{1}{x}.$$

This gives another explanation for formula (3.5) on p. 30.

**Example 22** Suppose we would like to find the slope of the line tangent to the circle centered at the origin with radius 5, at the point $(3,4)$. The equation of the circle is of course $x^2+y^2=25$. The hard way to do it is to first solve the equation for $y=\pm\sqrt{25-x^2}$. For the given point $(3,4)$, we choose the positive solution $y=\sqrt{25-x^2}$ and differentiate using the chain rule: $y'=\frac{1}{2}\left(25-x^2\right)^{-\frac{1}{2}}\cdot(-2x)=\frac{-x}{\sqrt{25-x^2}}$. Next, we evaluate the derivative $y'(3)=-\frac{3}{4}$.
One can avoid the square roots altogether by differentiating the implicit equation $2x+2yy'=0$. We solve easily for $y'=-\frac{x}{y}$ and at the given point get the same answer of $-\frac{3}{4}$.

## 3.4 Inverse trigonometric functions

Functions $f$ and $g$ are **inverses** of each other when

$$(g\circ f)(x) = (f\circ g)(x) = x$$

In other words, one function can undo what the other did. A function inverse to $f$ is denoted by $f^{-1}$, and this *has nothing to do* with the reciprocal. Yes, we are somewhat inconsistent: when we write $\sin^2 x$ we do mean $(\sin x)^2$, but when we write $\sin^{-1} x$, we mean the inverse *function* (see Section 1.7).

In order to have an inverse, the function must be one-to-one. That is, it must assign different values to different inputs. Otherwise, how could we expect the inverse function to guess the value of $x$, based solely on the value of $f(x)$? In other words, the graph of an invertible function must pass the "horizontal line test": any horizontal line crosses the graph at most once. A point $(a,b)$ lies on the graph of $f$ if and only if $(b,a)$ lies on the graph of $f^{-1}$.

The derivative of the inverse function $g$ of the function $f$ can be calculated using the formula

$$g'(x) = \frac{1}{f'(g(x))},$$

which can be obtained by considering the symmetry of the graphs of $f$ and $g$. We did exactly that in Section 3.1.2. However, it is usually easier to use implicit differentiation as follows. If $y = f^{-1}(x)$, then $x = f(y)$. Differentiating both sides (treating $x$ as the variable and $y$ as the function), we get:

$$1 = f'(y) \cdot y'.$$

Hence $y' = \frac{1}{f'(y)}$, where $y = f(x)$.

**Example 23** Suppose $y = \sin^{-1} x$. Find $y'$. We have

$$x = \sin y$$

with $-\frac{\pi}{2} \leq y \leq \frac{\pi}{2}$. Differentiating both sides with respect to $x$, we get

$$1 = \cos y \cdot y',$$

hence

$$y' = \frac{1}{\cos y}.$$

Using trigonometry, we can write $\cos y = \sqrt{1 - \sin^2 y} = \sqrt{1 - x^2}$. (We choose the positive square root, since we know that $y$ is in the first or fourth quadrant.) Therefore

$$\left(\sin^{-1} x\right)' = \frac{1}{\sqrt{1 - x^2}}.$$

In a similar fashion one can obtain:

$$\begin{aligned} \left(\cos^{-1} x\right)' &= \frac{-1}{\sqrt{1 - x^2}}, \\ \left(\tan^{-1} x\right)' &= \frac{1}{1 + x^2}. \end{aligned}$$

These formulas are good to remember, mostly because they are useful in integration. The function $y = \frac{1}{1+x^2}$ looks very innocent. It is eye-opening to realize that inverse trigonometric functions are needed to find an antiderivative of this function. We will examine these matters in more detail in Section 6.4.

## 3.5 Logarithmic differentiation

**Example 24** Find the derivative of

$$y=\sqrt{\frac{(2x+3)^{10}(x+5)}{(x-1)^{7}}}.$$

We could of course do it by "brute force," that is, apply the chain rule, then the quotient rule, then the chain rule again. Good luck! Instead, let us take a logarithm of both sides before taking the derivative, and use the properties of logarithms:

$$\begin{aligned}
\ln y &= \ln\left(\sqrt{\frac{(2x+3)^{10}(x+5)}{(x-1)^{7}}}\right)=\ln\left[\left(\frac{(2x+3)^{10}(x+5)}{(x-1)^{7}}\right)^{\frac{1}{2}}\right]\\
&= \frac{1}{2}\ln\left(\frac{(2x+3)^{10}(x+5)}{(x-1)^{7}}\right)\\
&= \frac{1}{2}\left[\ln\left((2x+3)^{10}(x+5)\right)-\ln\left((x-1)^{7}\right)\right]\\
&= \frac{1}{2}\left[\ln\left((2x+3)^{10}\right)+\ln(x+5)-\ln\left((x-1)^{7}\right)\right]\\
&= \frac{1}{2}\left[10\ln(2x+3)+\ln(x+5)-7\ln(x-1)\right]\\
&= 5\ln(2x+3)+\frac{1}{2}\ln(x+5)-\frac{7}{2}\ln(x-1).
\end{aligned}$$

Then take the derivative of both sides, treating $y$ as a function of $x$:

$$\frac{y'}{y}=5\frac{2}{2x+3}+\frac{1}{2}\frac{1}{x+5}-\frac{7}{2}\frac{1}{x-1},$$

and multiply both sides by $y$:

$$y'=\left(\frac{10}{2x+3}+\frac{1}{2}\frac{1}{x+5}-\frac{7}{2}\frac{1}{x-1}\right)\sqrt{\frac{(2x+3)^{10}(x+5)}{(x-1)^{7}}}.$$

(OK, so it does not look very pretty, but neither did the original function.)

The technique used in the example is called **logarithmic differentiation**. It is helpful when differentiating functions which are products or quotients of simpler expressions raised to various powers. It can also be used to quickly obtain the product rule:

**Example 25** Suppose $y = x^\alpha$. Then $\ln y = \ln(x^\alpha) = \alpha \ln(x)$ and after differentiating both sides we obtain

$$\frac{y'}{y} = \alpha \frac{1}{x}.$$

Multiplying both sides by $y$, we get the product rule (3.9).

For more examples, see problems 4 and 5.

## 3.6 Problems

1. Let $f(x) = x^4$. Explain why the function has a local (and global) minimum at $x = 0$, which is a critical point. What does the second derivative test tell us about $f$ at $x = 0$? Is the first derivative test more useful in this case?

2. Find the first and second derivatives of the functions

   (a) $y = \ln(1 + x^2)$,

   (b) $y = \arctan(x)$,

   (c) $y = \ln(\cos(3x))$,

   (d) $y = e^{x^2}$,

   (e) $y = e^{-x} \sin(2x)$.

3. Consider a function $y = xe^{-x}$, for $x \geq 0$. Find the minimum and maximum values of this function, the intervals of increase and decrease and the intervals of concavity. Use this information to sketch the graph of $y(x)$.

4. Use implicit differentiation to find the derivative of $y = \log_b x$ by rewriting it first as $b^y = x$.

5. Use logarithmic differentiation to differentiate the function $y = x^x$.

6. Verify that the function $y = a\sin(2t) + b\cos(2t)$ is a solution of the differential equation $y'' = 4y$, which models simple harmonic motion. Then find values of $a$ and $b$ such that $y(0) = 3$ and $y'(0) = 1$. Use a graphing device to graph $y(x)$ and interpret the graph as the position of a mass on a spring.

7*. A more realistic approach to the problem of a mass on a spring is to account for a damping force. This leads to a differential equation $y'' = -ky - dy'$.

   (a) Verify that a function $y_1 = e^{-t}\sin(2t)$ is a solution of the equation $y'' = -5y - 2y'$.

   (b) Check that the same is true about $y_2 = e^{-t}\cos(2t)$.

   (c) Show that if any two functions satisfy this differential equation, then their linear combination also does. Therefore, any linear combination $y(t) = ay_1(t) + by_2(t)$, where $a$ and $b$ are arbitrary constants is also a solution of the given differential equation.

   (d) Use a graphing device to graph the functions $y_1$, $y_2$ and some linear combination of them. Does the graph look plausible if interpreted as the position of a mass on a spring with damping?

8. The primary hyperbolic functions are defined as $\sinh(x) = \frac{e^x - e^{-x}}{2}$ and $\cosh(x) = \frac{e^x + e^{-x}}{2}$. Verify that

   (a) $\frac{d}{dx}[\sinh(x)] = \cosh(x)$;

   (b) $\frac{d}{dx}[\cosh(x)] = \sinh(x)$;

   (c) $\cosh^2(x) - \sinh^2(x) = 1$;

   (d) Derive the formula for the inverse hyperbolic function $\tanh^{-1}(x) = \frac{1}{2}\ln\left(\frac{1+x}{1-x}\right)$, where $\tanh(x) = \frac{\sinh(x)}{\cosh(x)}$.

   (e) Verify that $\frac{d}{dx}\left[\tanh^{-1}(x)\right] = \frac{1}{1-x^2}$.

# Chapter 4

# Some Applications of Derivatives

In this chapter a few selected mathematical applications of derivatives are discussed. In particular, we use derivatives to study asymptotic behavior of functions and develop techniques of finding limits. Also, some numerical methods for solving equations are mentioned. We briefly describe a few examples of optimization and related rates problems. However, our main effort is not directed toward solving a large number of real life problems. Instead, we try to develop mathematical intuition and tools which can be applied in various situations. In particular, the limits of functions at infinity are studied, not just for their own sake, but because they help us grasp the behavior of functions. This is not only a good exercise for the mind. The ability to reach qualitative understanding of the behavior of various mathematical expressions, by analyzing their formulas, is a skill useful in many different settings.

## 4.1 More about limits, l'Hôpital's rule

We begin by extending the definition of the limit to allow for the variable $x$ to approach infinity and for the limit to be infinite. The informal definition

9 in Chapter 2 can still be used, but we need to specify what it means for a number to approach $\infty$. For $x$ to be close to a number $a$ means that the distance between $x$ and $a$ is smaller than some small number: $|x-a|<\varepsilon$. In a similar fashion, $x$ is close to infinity if it is larger than some large number $M$. This leads to the following formalizations, allowing the variable $x$ or the values $f(x)$ to approach infinity.

**Definition 26** *We say that*

$$\lim_{x\to\infty} f(x) = L$$

*if for every $\varepsilon>0$, there exists an $M>0$ such that $|f(x)-L|<\varepsilon$ whenever $x>M$.*

**Definition 27** *We say that*

$$\lim_{x\to a} f(x) = \infty$$

*if for every $M>0$ there exists an $\delta>0$ such that $f(x)>M$ whenever $0<|x-a|<\delta$.*

**Exercise 28** *State the exact meaning of the intuitively clear statements $\lim_{x\to a} f(x) = -\infty$ and $\lim_{x\to\infty} f(x) = \infty$.*

Definition 26 describes an aspect of the long-term behavior of a function, which is the value to which $f(x)$ is close, for sufficiently large $x$. Geometrically, it means that the graph of the function $f$ has a horizontal asymptote $y=L$. On the other hand, definition 27 applies to a local "explosion" of values of $f$ when $x$ is near $a$. In this case we say that $f$ has a vertical asymptote $x=a$. (If we restrict our attention to the one-sided limit, say $\lim_{x\to a^+} f(x)$, then we say that $f$ has a one-sided asymptote.)

Let us recall that in order to develop formulas (3.6) on page 31 it was necessary to find the limit $\lim_{x\to 0}\frac{\sin(x)}{x}=1$. In the fraction $\frac{\sin(x)}{x}$ both the numerator and the denominator approach 0 as $x$ gets close to 0. In general, it is not possible to draw any conclusions about the limit of the form $\lim_{x\to a}\frac{f(x)}{g(x)}$, based solely on the fact that both the numerator and the denominator approach zero, that is, $\lim_{x\to a} f(x) = \lim_{x\to a} g(x) = 0$. The critical issue is which of the functions approaches zero faster.

We are faced with the same difficulty when $\lim_{x\to a} f(x) = \lim_{x\to a} g(x) = \infty$. For instance, we have $\lim_{x\to\infty} \frac{x}{x^2} = 0$, $\lim_{x\to\infty} \frac{x^2}{x} = \infty$ and $\lim_{x\to\infty} \frac{3x}{x} = 3$. Again, it is important which of the functions grows faster, and how much faster.

A commonly accepted shorthand to describe this difficulty is to say that the expressions $\frac{\infty}{\infty}$ and $\frac{0}{0}$ are **indeterminate symbols** or **indeterminate expressions**. It means that one should not jump to any conclusions about this limit simply because $f(x)$ and $g(x)$ both approach 0 (or both approach $\infty$) when $x \to a$. It does not mean that the limit $\lim_{x\to a} \frac{f(x)}{g(x)}$ does not exist, just that there is more work to be done. Some other indeterminate expressions, such as $\infty - \infty$, $0 \cdot \infty$, $0^0$ and $1^\infty$, are illustrated in the examples.

**Example 29** Consider the limit $L = \lim_{x\to\infty} \left(\sqrt{x+1} - \sqrt{x}\right)$. Since both $\sqrt{x}$ and $\sqrt{x+1}$ approach $\infty$ as $x \to \infty$, we cannot draw any conclusion about their difference. However, $\sqrt{x+1} - \sqrt{x} = \frac{\left(\sqrt{x+1}-\sqrt{x}\right)\left(\sqrt{x+1}+\sqrt{x}\right)}{\sqrt{x+1}+\sqrt{x}} = \frac{x+1-x}{\sqrt{x+1}+\sqrt{x}} = \frac{1}{\sqrt{x+1}+\sqrt{x}}$. Therefore $L = \lim_{x\to\infty} \frac{1}{\sqrt{x+1}+\sqrt{x}} = 0$.

**Example 30** Let $L = \lim_{x\to 0^+} (x \cot x)$. Clearly $x \to 0$ and $\cot x \to \infty$ when $x \to 0$ from the right. Yes, it is true that zero times anything is zero. However, in the expression $x \cot x$, while the first term "$x$" *approaches* zero, it is *not actually equal* to zero. As $x \to 0$, one of the factors in the expression $x \cot x$ gets closer and closer to zero, while the other one gets larger and larger. The resulting product could be large or small, depending on which of the factors prevails. In this case one can notice that $x \cot x = x \frac{\cos x}{\sin x} = \left(\frac{x}{\sin x}\right) \cdot \cos x$. Since $\lim_{x\to 0} \frac{x}{\sin x} = 1$ and $\lim_{x\to 0} (\cos x) = 1$, the limit $L = 1 \cdot 1 = 1$.

One way to approach the problem of finding limits of the type $\frac{0}{0}$ or $\frac{\infty}{\infty}$ is to find the limit of the ratio of derivatives instead:

**Theorem 31** *(l'Hôpital's rule) Suppose that* $\lim_{x\to a} f(x) = \lim_{x\to a} g(x) = 0$ *(or* $\infty$*). Then*

$$\lim_{x\to a} \frac{f(x)}{g(x)} = \lim_{x\to a} \frac{f'(x)}{g'(x)},$$

*provided that the second limit exists.*

This theorem is actually quite delicate. In lieu of a formal proof imagine two starships not too far from our planet. Let $f$ and $g$ denote the corresponding distances between the earth and the starships. Suppose that the starships are moving away from the earth and that the speed of the first one is eventually about twice the speed of the second one: $f' \approxeq 2g'$. Eventually, the first starship should be about twice as far from the earth as the second one: $f \approxeq 2g$, regardless of their initial positions and maneuvers. This is what l'Hôpital's rule is saying in the case $a = \infty$ and $f, g \to \infty$ as $x \to \infty$.

**Example 32** Let us use l'Hôpital's rule to find $\lim_{x\to 0} \frac{\sin x}{x}$. We know of course that the answer is 1, and we used this fact to establish the formulas for derivatives of sine and cosine in Section 3.1.3. We have $\lim_{x\to 0} \sin x = \lim_{x\to 0} x = 0$. Therefore, l'Hôpital's rule can be applied. We get $\lim_{x\to 0} \frac{\sin x}{x} = \lim_{x\to 0} \frac{\cos x}{1} = 1$. (Yes, this is an example of a circular reasoning, since we needed to know the value of the limit in question in order to see that $(\sin x)' = \cos x$. Still, it all fits together.)

**Example 33** Consider $\lim_{x\to 0^+} x \ln(x)$. As $x \to 0^+$, the other factor $\ln x \to -\infty$. In order to apply l'Hôpital's rule, one needs first to rewrite the expression $x \ln(x)$ as a quotient. This is fairly simple, as $a \cdot b = \frac{b}{1/a}$. In this case $x \ln(x) = \frac{\ln x}{1/x}$. We have

$$\begin{aligned} \lim_{x\to 0+} x \ln(x) &= \lim_{x\to 0+} \frac{\ln(x)}{1/x} = \lim_{x\to 0+} \frac{1/x}{-1/x^2} \\ &= \lim_{x\to 0+} \frac{1/x}{-1/x^2} \cdot \frac{x^2}{x^2} = \lim_{x\to 0+} -\frac{x}{1} = 0. \end{aligned}$$

One is frequently tempted to obtain the same answer by a careless argument of the kind: zero times anything is zero. Again, $x$ is not 0, but only close to it. The limit actually is 0 because as $x$ steadily approaches 0, it prevails over $\ln x$ which "crawls" to $\infty$.

**Example 34** Consider an expression $E(x) = (1 + x)^{\frac{1}{x}}$ for small positive values of $x$. The expression $1 + x$ is close to 1. We know that 1 raised to any power is also 1. The problem is that $1 + x \neq 1$, *it is only close to* 1. As $x$ approaches 0 (let us say from the positive side), the exponent $\frac{1}{x} \to \infty$. If the reader recalls problem 10 from Chapter 1, by numerical experimentation we can see that $\lim_{x\to 0} E(x) \approxeq 2.718$. In order to see how one can use l'Hôpital's rule in this setting, we need to learn another trick. Consider $H(x) = \ln E(x)$. Suppose we know that $H(x) \to L$ as $x \to 0$. Since $E(x) = e^{H(x)}$, continuity of the exponential function implies that

$$\lim_{x\to 0} E(x) = \lim_{x\to 0} e^{H(x)} = e^{\lim_{x\to 0} H(x)} = e^{L}.$$

This gives us a method of finding the desired limit, as long as we can find $\lim_{x\to 0} H(x)$. In other words, we replace the original limit of $E(x)$ with a new one: $L = \lim_{x\to 0} \ln(E(x))$, which may be easier to find. The answer to the original problem is $e^L$. We have

$$H(x) = \ln\left[(1+x)^{\frac{1}{x}}\right] = \frac{1}{x}\ln(1+x) = \frac{\ln(1+x)}{x};$$

therefore by l'Hôpital's rule we have

$$\lim_{x\to 0} H(x) = \lim_{x\to 0} \frac{\ln(1+x)}{x} = \lim_{x\to 0} \frac{\frac{1}{1+x}}{1} = 1.$$

Hence

$$\lim_{x\to 0} (1+x)^{\frac{1}{x}} = e^1 = e.$$

**Example 35** Along the same lines as in the previous example, let us examine the behavior of the function $f(x) = x^{2x}$ as $x \to 0^+$. Zero raised to any positive power is still zero. On the other hand, any positive number raised to the power of zero is one. Again, this apparent contradiction is clarified as soon as we realize that while $x$ is close to 0, is does not equal 0. Let us use the logarithm trick we just learned and first find $\lim_{x\to 0+} \ln\left(x^{2x}\right) = \lim_{x\to 0+} 2x\ln(x) = 0$, as we know from Example 33. Therefore

$$\lim_{x\to 0+} x^{2x} = e^{\lim_{x\to 0+}\left(\ln x^{2x}\right)} = e^0 = 1.$$

### 4.1.1 Rate of growth of functions

It was apparent in several examples, that it is not important that function $f(x) \to \infty$ (or 0) as $x \to a$, but also how fast this happens. It is helpful to be aware of the various rates at which functions may grow as $x \to \infty$. The following is a list of certain types of functions, in the order of decreasing rate of growth.

1. Exponential functions $y = b^x$, where $b > 1$, approach $\infty$ very fast as $x \to \infty$. The rate of growth is faster for larger values of $b$. For example, for sufficiently large $x$, the value of $3^x$ is several times larger than $e^x$.

2. Power functions $y = x^\alpha$, where $\alpha > 0$. This means, in particular, that for a polynomial $P_n(x)$ of degree $n$, the behavior of its leading term $a_n x^n$ is the only thing that matters for sufficiently large values of $x$. Therefore, for two polynomials $P_n(x) = a_n x^n + a_{n-1}x^{n-1} + \cdots + a_1 x + a_0$ and $Q_m(x) = b_m x^m + b_{m-1}x^{m-1} + \cdots + b_1 x + b_0$, of degree $n$ and $m$, accordingly, we have $\frac{P_n(x)}{Q_m(x)} \approxeq \frac{a_n x^n}{b_m x^m}$, for large values of $x$. Therefore:
$$\lim_{x\to\infty} \frac{P_n(x)}{Q_m(x)} = \begin{cases} 0 & \text{if } n < m \\ \frac{a_n}{b_n} & \text{if } n = m \\ \operatorname{sgn}\left(\frac{a_n}{bm}\right) \cdot \infty & \text{if } n > m \end{cases},$$
(the last case means $\pm\infty$, where the sign of $\infty$ is the same as the sign of the ratio of the leading coefficients $\frac{a_n}{bm}$).

3. Logarithmic functions $y = \log_b x$, where $b > 1$. These functions approach $\infty$ very slowly. For instance, for the decimal logarithm to reach the value 66, we need $x = 10^{66}$, an approximate number of atoms in the galaxy. Clearly, for larger values of the base $b$, the growth of $\log_b x$ is a little faster.

4. If someone likes to take things really slow, a function $y = \log(\log x)$ should be quite satisfying. (This function occurs frequently in number theory, but that is a whole other story.)

5. Constant and bounded functions, such as $y = \sin x$ or $y = 7\cos(5x)$, do not approach $\infty$ at all.

**Example 36** Using rates of growth of functions we can immediately conclude that $\lim_{x\to\infty} \frac{\ln x}{x} = 0$, $\lim_{x\to\infty} \frac{e^x}{-x^3-3x^2+5} = \infty$, $\lim_{x\to\infty} \frac{\cos x}{\ln x} = 0$, but $\lim_{x\to\infty} \frac{\ln x}{\cos x}$ does not exist, as the denominator keeps changing its sign.

**Example 37** $\lim_{x\to\infty} \frac{2^x+x^2-3}{3^x+x^5-7} = \lim_{x\to\infty} \frac{2^x}{3^x} = 0.$

**Example 38** Since the logarithm function is continuous, we have
$\lim_{x\to\infty} \left[\ln\left(\frac{3x^2-x}{ex^2-7}\right)\right] = \ln\left[\lim_{x\to\infty}\left(\frac{3x^2-x}{ex^2-7}\right)\right] = \ln\left(\frac{3}{e}\right) = \ln 3 - \ln e = \ln 3 - 1.$

## 4.2 Optimization

Calculus is frequently used to minimize certain quantities, such as cost, given a set of constraints we have to preserve. We might also wish to maximize some quantity, such as profit, production, area, volume, etc., given certain restrictions. The good news is, that a continuous function on a closed interval $[a, b]$ always attains the maximum (minimum) value. This is known as the **extreme value theorem** (or **Weierstrass** theorem). The statement is easy to believe, but more difficult to prove.

It is natural to consider derivatives in such problems, as extreme values of a differentiable function can only be attained at stationary points ($f'(x) = 0$) or at the edge of the domain of $f$. There is no general approach which would solve all problems of this kind. Typically, it helps to draw some kind of a diagram and to introduce appropriate notation. It is crucial to set up the constraints and the objective function correctly. The constraints can be used to express the objective function in terms of a single variable, at which point we can employ derivatives to find the optimal value of the objective function.

**Example 39** What is the largest value a product of two positive numbers can be, given that their sum does not exceed 10? In other words, we wish to maximize the product $xy$, subject to the constraint $x + y \leq 10$. It is apparent that we can assume that $x + y = 10$, or $y = 10 - x$. The quantity to be minimized can be written as a function of just one variable $f(x) = x(10 - x) = -x^2 + 10x$. We have $f'(x) = -2x + 10$ and $f''(x) = -2$. Therefore, $f$ is concave down at all times and it has its only maximum point at $x = 5$. The maximum value is $f(5) = 25$. We could have also noticed that the graph of $f$ is a parabola opening down, whose $x$-intercepts are 0 and 10. The vertex is at the value of $x = \frac{0+10}{2} = 5$.

**Example 40** What is the largest area of a rectangular garden plot which can be fenced using 20 feet of fence? If the dimensions of the plot are $x$ and $y$, the constraint is $2x + 2y \leq 20$, which is exactly what we had in the previous example.

**Example 41** We would like to make a cylindrical uncovered barrel. The cost of the material for the bottom is $6 per square foot and just $3 per square foot for the rest of it. What is the most economical shape of the barrel?

Let us denote the radius of the barrel by $r$ and its height by $h$. The volume $V = \pi r^2 h$ and the cost $C = 6\pi r^2 + 3 \cdot 2\pi r h$. We can fix the desired volume of the barrel and minimize the cost of the material or fix the cost of the material and maximize the volume.
Method I: Suppose that the volume is fixed to be $V$. We can use the constraint $V = \pi r^2 h$ to express $h$ in terms of $r$, that is $h = \frac{V}{\pi r^2}$. The cost $C$ can now be written as a function of $r$ :

$$C(r) = 6\pi r^2 + 3 \cdot 2\pi r \frac{V}{\pi r^2} = 6\pi r^2 + \frac{6V}{r}.$$

To find stationary points, we set

$$C'(r) = 12\pi r - \frac{6V}{r^2} = 0,$$

which leads to $r^3 = \frac{V}{2\pi}$. In other words, we need $r^3 = \frac{\pi r^2 h}{2\pi}$ or $h = 2r$. The height of the barrel should be the same as its diameter. We can be sure to have minimized the cost $C$, by checking that $C''(r) = 12\pi + \frac{12V}{r^3} > 0$ for all $r > 0$, so the function $C(r)$ is concave up.
Method II: Now suppose that we have exactly $C$ dollars available and we would like to build the largest possible container. This time $C = 6\pi r^2 + 3 \cdot 2\pi r h$ is the constraint, which we can use to say that $h = \frac{C - 6\pi r^2}{6\pi r}$. Substituting this into the formula for the volume, we get

$$V(r) = \pi r^2 \frac{C - 6\pi r^2}{6\pi r} = \frac{Cr}{6} - \pi r^3.$$

To find the stationary point, we set

$$V'(r) = \frac{C}{6} - 3\pi r^2 = 0$$

to obtain $C = 18\pi r^2$ or $r = \sqrt{\frac{C}{18\pi}}$. Again, $V''(r) = -6\pi r < 0$, so the function $V(r)$ is concave down and has the global maximum at its only stationary point. The dimensions of the optimal barrel follow since $h = \frac{C-6\pi r^2}{6\pi r} = \frac{18\pi r^2 - 6\pi r^2}{6\pi r} = 2r$.

**Example 42** Find the points on the parabola $y = x^2$ which are closest to the point $(0,2)$. The distance from the point on the line $(x, x^2)$ to the point $(0,2)$ is $d(x) = \sqrt{x^2 + (x^2-2)^2}$. However, instead of minimizing $d(x)$ it is simpler to minimize its square $f(x) = [d(x)]^2 = x^2 + (x^2-2)^2$. We have $d'(x) = 2x - 2(x^2-2)2x = 2x\left[1 - 2(x^2-2)\right] = 2x(5-2x^2)$. Solving the equation $d'(x) = 0$ we obtain $x = 0$ and $x = \pm\sqrt{5/2} = \pm\sqrt{10}/2 \approxeq \pm 1.58$. By inspecting the sign of $d'$ we can see that $d$ has a local maximum at $x = 0$ and two local minima at $x = \pm\sqrt{5/2}$. The points on the parabola closest to $(0,2)$ are $\left(\sqrt{5/2}, 5/2\right)$ and $\left(-\sqrt{5/2}, 5/2\right)$.

## 4.3 Related rates

Another application of derivatives occurs naturally when several quantities are changing in time. If these quantities are related to each other, then the rate of change one quantity may determine the rate of change of the other. The most important thing to keep in mind is to keep variables variable and not to plug in specific values too early in the problem.

**Example 43** A classical example is that of a person walking away from a lamppost; let us say the lamppost is 8 feet tall, and that the person's height is 6 feet. How quickly is the length of the shadow changing when the person is 10 feet away from the lamppost, walking away at 7 feet per second? Denote the distance of the person from the lamppost by $x(t)$ and the length of the shadow by $y(t)$. Considering similar triangles, we obtain

$$\frac{x+y}{8} = \frac{y}{6},$$

which simplifies to $6(x+y) = 8y$, or $y = 3x$. Since we know that $x'(t) = 7$ at all times, the answer is $y'(t) = 3 \cdot x'(t) = 21$. In other words, the length of the shadow changes three times as fast as the distance from the lamppost.

A typical mistake in solving related rates problems is to replace variables with constants before differentiating. To illustrate this, suppose that instead of $x(t)$ we wrote 10. The equation would be $\frac{10+y}{8} = \frac{y}{6}$, which we can solve for $y = 30$. It is true that the shadow is 30 feet long when the person is 10 feet away from the lamppost, but this does not help us find any derivatives.

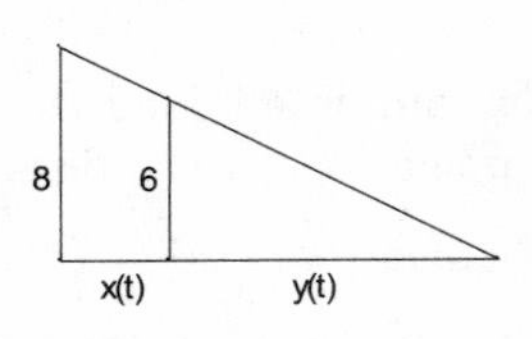


**Example 44** Let us revisit Example 3 from Section 3.3. This time, however, let us assume that the point is moving on the circle $x^2 + y^2 = 25$, that is, $x$ and $y$ are functions of time $t$. One can think of this as taking a ride on a Ferris wheel of radius 5 ft. (a rather small one). Suppose that the wheel is turning at a constant angular velocity and that a full turn takes 30 seconds. It is natural to ask how fast the point is rising when its coordinates are $(3, 4)$, for example. One possible parametrization is $x(t) = 5\cos\left(\frac{2\pi}{30}t\right)$ and $y(t) = 5\sin\left(\frac{2\pi}{30}t\right)$. We have $y'(t) = 5\cos\left(\frac{2\pi}{30}t\right) \cdot \frac{2\pi}{30} = x(t) \cdot \frac{2\pi}{30}$. At the point in question, $y' = 3 \cdot \frac{2\pi}{30} \approxeq 0.628\,32$ ft./sec. Again, it was important to keep both functions as functions rather than to replace them with specific values for taking derivatives.

## 4.4 Function approximation, Taylor polynomials

Albert Einstein used to say that *"Things should be made as simple as possible, but not simpler."* It is often useful to approximate a complicated function by a simpler one. When we do that, we get a simpler formula to work with, at a price of losing some information about the original function.

The crudest approximation of a function $f$ near $x = x_0$ is by a constant function $P_0(x) = f(x_0)$. There is only one thing good about this approximation: the values of $f$ and $P_0$ agree at $x = x_0$.

One step up from this is the linear approximation, which is an approximation by the straight line tangent to the graph of $f$ at the point $(x_0, f(x_0))$.

The tangent line is given by a polynomial of degree one:

$$P_1(x) = f(x_0) + f'(x_0) \cdot (x - x_0),$$

also known as the point-slope form of the line equation (compare with equation (2.4) in Section 2.2). This time, the values of the function $f$ and its derivative $f'$ are matched by the corresponding values of $P_1$ and $P_1'$ at $x = x_0$.

It is natural to expect that, if we allow the approximating polynomial $P_2$ to be quadratic, then we should also be able to match the second derivative. Allowing a cubic should give us enough flexibility to match the third derivative, and so on. As a result, we can find polynomials, that fit the graph of $f$ more snugly near a selected value of $x = x_0$.

For simplicity, we will consider approximating functions near $x_0 = 0$. Specifically, we would like to find a cubic polynomial $P_3(x) = a_0 + a_1 x + a_2 x^2 + a_3 x^3$, such that $P_3(0) = f(0)$, $P_3'(0) = f'(0)$, $P_3''(0) = f''(0)$ and $P_3'''(0) = f'''(0)$. We differentiate and plug in $x = 0$ to get

$$\begin{array}{ll}
P_3(x) = a_0 + a_1 x + a_2 x^2 + a_3 x^3, & P_3(0) = a_0, \\
P_3'(x) = a_1 + 2a_2 x + 3a_3 x^2, & P_3'(0) = a_1, \\
P_3''(x) = 2a_2 + 3 \cdot 2a_3 x, & P_3''(0) = 2a_2, \\
P_3'''(x) = 3 \cdot 2a_3, & P_3''(0) = 3 \cdot 2a_3.
\end{array}$$

Therefore, to match the corresponding values of the derivatives, we must take $a_0 = f(0)$, $a_1 = f'(0)$, $a_2 = \frac{f''(0)}{2}$, $a_3 = \frac{f'''(0)}{3!}$, which gives

$$P_3(x) = f(0) + f'(0) \cdot x + \frac{f''(0)}{2} \cdot x^2 + \frac{f'''(0)}{3!} \cdot x^3. \tag{4.1}$$

The same reasoning can be applied to a polynomial of degree $n$, that approximates a given function $f$ near an arbitrarily chosen $x_0$, which leads to the so-called **Taylor polynomial**

$$\begin{aligned}
P_n(x) &= f(x_0) + f'(x_0)(x - x_0) + \cdots + \frac{f^{(n)}(x_0)}{n!}(x - x_0)^n \\
&= \sum_{k=0}^{n} \frac{f^{(k)}(x_0)}{k!}(x - x_0)^k.
\end{aligned} \tag{4.2}$$

The special case of $x_0 = 0$ is called the **Maclaurin polynomial**.

**Example 45** Let us find third and fifth order Maclaurin polynomials for $f(x) = \sin x$. We have $f'(x) = \cos x$, $f''(x) = -\sin x$, $f'''(x) = -\cos x$, $f^{(4)}(x) = \sin x$ and $f^{(5)}(x) = \cos x$. Since $x_0 = 0$, we compute $f(0) = 0$, $f'(0) = 1$, $f''(0) = 0$, $f'''(0) = -1$, $f^{(4)}(0) = 0$ and $f^{(5)}(0) = 1$. Hence, by formula (4.1), we have

$$\begin{aligned} P_3(x) &= 0 + 1 \cdot x + \frac{0}{2} \cdot x^2 + \frac{-1}{3!} \cdot x^3 \\ &= x - \frac{1}{6}x^3. \end{aligned}$$

In the same fashion the fifth order polynomial just has two more terms:

$$\begin{aligned} P_5(x) &= P_3(x) + \frac{f^{(4)}(0)}{4!} \cdot x^4 + \frac{f^{(5)}(0)}{5!} \cdot x^5 \\ &= x - \frac{1}{6}x^3 + \frac{1}{120}x^5. \end{aligned}$$

We encourage the reader to graph the function $f(x) = \sin x$ and a few approximations by Maclaurin polynomials in the same window on a calculator.

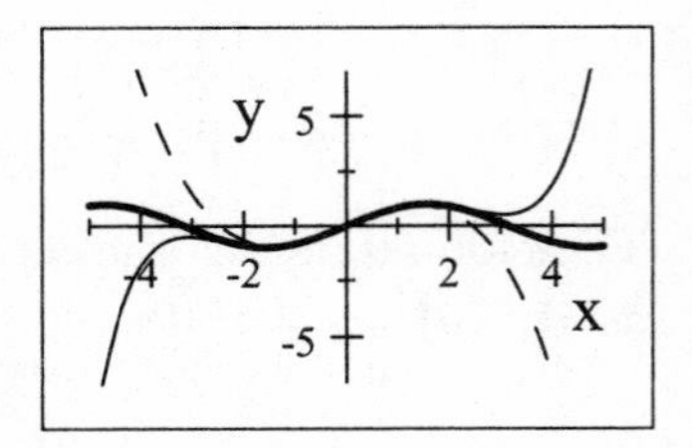


After the original function is replaced by its approximation, matters became simpler, but at a price: using the approximation instead of the actual function generates an error. It is natural to ask how big could this error be? We have the following error bound for the Taylor polynomial on a given interval $[a, b]$:

$$|f(x) - P_n(x)| \leq \frac{K_{n+1}}{(n+1)!} |x - x_0|^{n+1}, \tag{4.3}$$

where the constant $K_{n+1}$ is such that $\left|f^{(n+1)}(x)\right| \leq K_{n+1}$ for all values of $x \in [a, b]$. In other words, the error committed by using the $n$-th order Taylor polynomial can be estimated by looking at the $(n+1)$-st derivative of $f$. We need to keep in mind that it is not always necessary to get the lowest possible value for $K_{n+1}$, just a reasonable one. A proof of this error bound for $n = 2$ is included in Appendix A.2.

**Example 46** Consider Maclaurin polynomial approximations of $g(x) = e^x$ on the interval $[-5, 5]$. How do we find an $n$ large enough to guarantee that the approximation error does not exceed $0.01$? First of all, let us notice that all derivatives of $g$ are the same, $g^{(k)}(x) = e^x$, which means $g^{(k)}(0) = e^0 = 1$ for all values of $k$. We have

$$P_n(x) = \sum_{k=0}^{n} \frac{g^{(k)}(0)}{k!} x^k = \sum_{k=0}^{n} \frac{x^k}{k!} = 1 + x + \frac{x^2}{2} + \cdots + \frac{x^n}{n!},$$

which is nice to know, but not directly relevant to our question. What we need to look at is the error bound (4.3). To find the constant $K_{n+1}$, we notice that $\left|g^{(n+1)}(x)\right| = e^x$, which is an increasing function. Its largest value on the interval $[-5, 5]$ is $g(5) = e^5 \approxeq 148.41 < 150$. We can take $K_{n+1} = 150$. It is sufficient to take $n$ large enough so that

$$|e^x - P_n(x)| \leq \frac{150}{(n+1)!} |x|^{n+1} \leq 0.01.$$

The largest $|x|^{n+1}$ can be is $5^{n+1}$. By numerical experimentation we can see that $n = 19$ is sufficient to guarantee that $\frac{150}{(n+1)!} 5^{n+1} \leq 0.01$.

We might expect that by allowing higher order approximating polynomials, we can make the error smaller. Unfortunately, this does not always happen. The following example illustrates the concept of the **interval of convergence**.

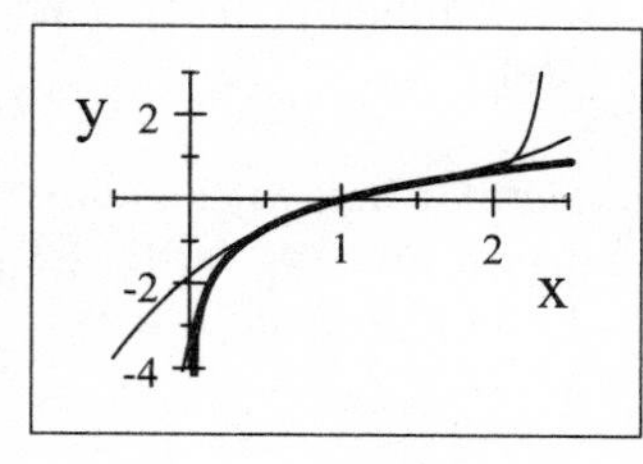


**Example 47** Let us approximate $h(x) = \ln x$ near $x_0 = 1$. We have $h'(x) = \frac{1}{x}$, $h''(x) = \frac{-1}{x^2}$ and so on. It is simple to verify that $P_2(x) = (x-1) - \frac{1}{2}(x-1)^2 + \frac{1}{3}(x-1)^3$ and in general $P_n(x) = \sum_{k=1}^{n} \frac{(-1)^{k+1}}{k}(x-1)^k$. Again, we can graph several of these polynomials ($P_3$ and $P_{19}$ are in the figure). We observe that for all $x \in (0, 2)$ the approximations are getting better with increasing values of $n$. However, for other values of $x$, the polynomials do not get better at all. We will come back to this topic in Chapter 8.

To finish this chapter on a positive note, let me assure that reader that the approximations for $e^x$ do get better with larger values of $n$. In fact, we may wish to add infinitely many terms and actually represent $e^x$ as an infinite "polynomial." Yes, this was a commercial for infinite series. If it sounds intriguing, there is no harm in reading the first page or two of Chapter 8 right away.

## 4.5 Problems

1. Find the limit and justify your answer: $\lim_{x \to 0^+} (1+2x)^{\frac{1}{x}}$. (Use the method outlined in Example 34.)

2. Find the limit $\lim_{x \to \infty} (2^x + 3^x)^{\frac{1}{x}}$. Hint: The answer is less obvious this time; it is no longer just $\lim_{x \to \infty} (b^x)^{\frac{1}{x}} = b$. The logarithmic trick is needed again to find the answer.

3. Make a convincing argument that $\lim_{x \to \infty} \frac{\cos x}{x} = 0$ by noticing that $-\frac{1}{x} \leq \frac{\cos x}{x} \leq \frac{1}{x}$.

4*. A right triangle has legs on the positive $x$- and $y$- axes, and its hypotenuse passes through the point $(4,1)$. Which such triangle has the smallest area? (There are several possible approaches to this problem. It is not very hard, but for some reason many students have difficulty setting it up or tend to get lost in the details, so just be careful.)

5. Find the dimensions of the most economical rectangular box with square base if the cost of the material for the walls, bottom and top is \$1, \$2 and \$3 per square foot, respectively. The desired volume of the box is a constant $V$.

6. A 10 ft. ladder leans against a wall. Suppose that the base of the ladder slides away from the wall. The base is moving away at 1 ft./sec. when it is 2 ft. away from the wall. Find the speed with which the top of the ladder is sliding down the wall at that moment.

7. Find Maclaurin polynomials of order 9 for $f(x) = \sin x$ and $g(x) = \cos x$. Note, that (about) half of the coefficients are zero in each of the polynomials.

8. Find the third order Taylor polynomial centered at $x_0 = 1$ for $g(x) = x^3$. Then multiply out your answer. Explain what you see.

9. (**Newton's method**) Solving an algebraic equation is equivalent to finding $x$-intercepts of an appropriate function. For example, it is easy to see that $x^3 - x = 0$ for $x = 0, 1$ or $-1$. Let us pretend we do know it and start by making a guess "solution" $x_0 = 2$.

   (a) Find an equation of the line tangent to the graph of $f(x) = x^3 - x$ at $x_0 = 2$.

   (b) Find the $x$-intercept of this tangent line and call it $x_1$.

   (c) Repeat this procedure to obtain $x_2$ and $x_3$. Observe that they are approaching 1.

   (d) Show that in general, the next iteration in the Newton's method is given by

$$x_{n+1} = x_n - \frac{f(x_n)}{f'(x_n)}.$$

   (e) Repeat this exercise, but start with different values of $x_0$. What do you observe?

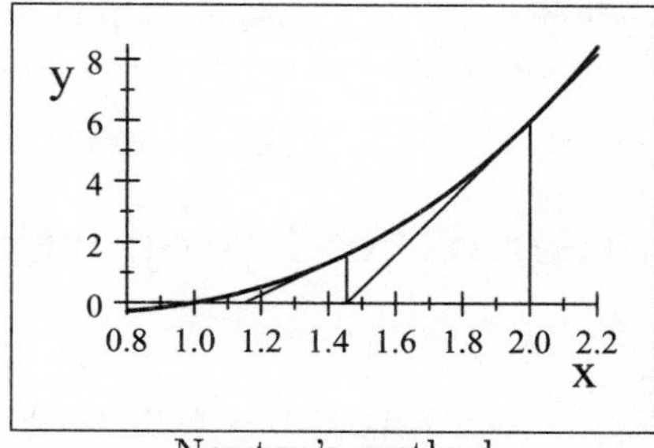


Newton's method

# Chapter 5

# The Integral and its Properties

The word "integral" is used to describe two different concepts: a definite integral is a number, while an indefinite integral is a function or, more precisely, a family of functions. The two concepts are very closely related. In order to integrate, we also need to develop some techniques for finding antiderivatives. Learning how to differentiate was complicated enough, but at least there were rules we could follow. To reverse the process of differentiation is significantly more difficult. In many cases there are no rules we could apply - just a bag of tricks, some of which might be useful. In this chapter we concentrate mostly on the concepts, while in the next one several techniques of integration are discussed.

## 5.1 Areas, integrals and properties of integration

We begin by taking an informal approach to the definition of **definite integral**. For a given function the definite integral over $[a, b]$ is a single number. Specifically, the integral $\int_a^b f$, also denoted by $\int_a^b f(x)\, dx$, is the signed area between the graph of $y = f(x)$ and the interval $[a, b]$.

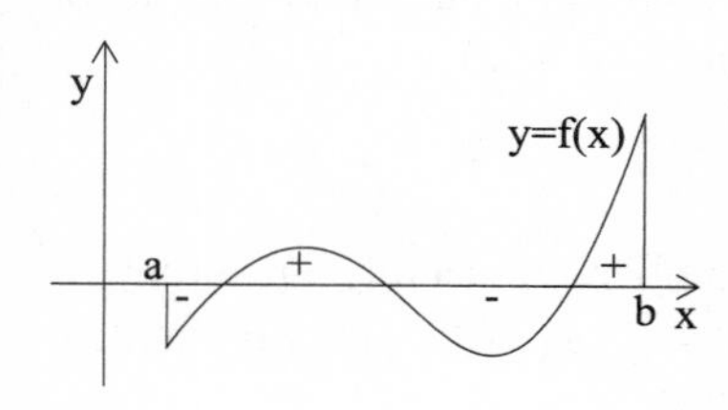


This means that whenever the graph of $f$ is above the $x$-axis, the area is counted as positive, and whenever the graph is below the $x$-axis, the area is counted as negative. What exactly do we mean by the area? We rely on an intuitive understanding of the concept of the area of a region. A formal definition of the integral will be given in Section 5.4, as a limit of a sequence of approximations.

**Example 48** We have $\int_{-1}^{5} 4dx = 4{\cdot}6 = 24$. This is because the function $f(x) = 4$ and the region under the graph of $f$ is a rectangle with base 6 and height 4.

**Example 49** To find $I = \int_{-2}^{2} \sqrt{4 - x^2}dx$, we observe that the graph of the function $f(x) = \sqrt{4 - x^2}$ is the upper half of the circle centered at the origin with radius $r = 2$. Therefore $I = \frac{1}{2}\pi r^2 = 2\pi$. If the interval of integration had not been chosen so nicely, the area would be much harder to calculate (see Section 6.4).

### Linearity

In general, for any constant $c$ and any integrable functions $f(x)$ and $g(x)$ we have

$$\int_a^b f(x) + g(x)\,dx = \int_a^b f(x)\,dx + \int_a^b g(x)\,dx, \qquad (5.1)$$

$$\int_a^b c \cdot f(x)\,dx = c \cdot \int_a^b f(x)\,dx.$$

That is, an integral of a sum of functions is the sum of their integrals and the constants can be "pulled out." This makes intuitive sense. For a given function $f(x)$, the graph of $y = 5f(x)$ can be obtained from the graph of $y = f(x)$ by stretching it vertically by a factor of 5. The area of the new region below the graph of $y = 5f(x)$ over a given interval $[a, b]$ is five times larger than the original area. In the language of integrals, $\int_a^b 5f(x)\,dx = 5 \cdot \int_a^b f(x)\,dx$. A similar argument can be made for a sum of two functions, as the $y$-coordinates of all points on the graph of $y = f(x) + g(x)$ are sums of the $y$-coordinates of the points on the original graphs of $y = f(x)$ and $y = g(x)$.

### Monotonicity

An integral of a larger function is a larger number. Formally speaking, if $f(x) \le g(x)$ for all $x \in [a,b]$, then $\int_a^b f(x)\,dx \le \int_a^b g(x)\,dx$. This inequality makes geometric sense since the graph of the function $f(x)$ lies below the graph $g(x)$. As a special case, suppose that for some constants $m$ and $M$ we have $m \le f(x) \le M$, for all $x \in [a,b]$. The region below the graph of $f$ is "boxed in" by the larger rectangle with height $M$. On the other hand, the small rectangle, with height $m$, fits under the graph of $f$. We have

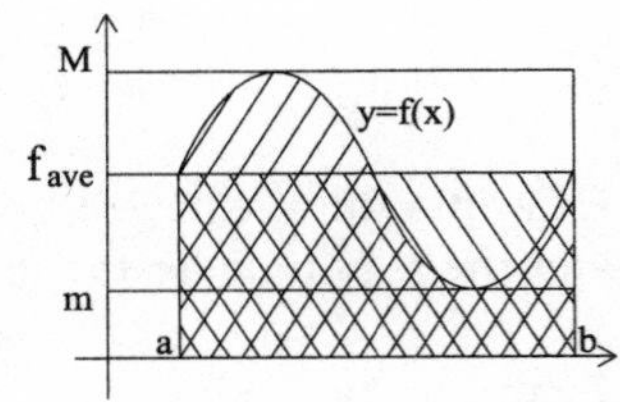


$$m(b-a) = \int_a^b m dx \le \int_a^b f(x)\,dx \le \int_a^b M dx = M(b-a). \tag{5.2}$$

**Example 50** To show that the integral $0 \le \int_0^3 \left|\sin\left(x^2\right)\right| dx \le 3$, it is enough to notice the the function $f(x) = \left|\sin\left(x^2\right)\right|$ takes values between 0 and 1. Estimates like this are useful when one cannot find the exact value of the integral.

### Average value of a function $f$

The average value of the function $f$ over the interval $[a,b]$ is defined as

$$f_{AVE} = \frac{\int_a^b f(x)\,dx}{b-a}. \tag{5.3}$$

To see that this is a sensible definition, notice that if the values of $f$ stay between $m$ and $M$, then the inequality (5.2) holds. Surely there must be a number $f_{AVE}$, such that the area below the horizontal line $y = f_{AVE}$ over the interval $[a,b]$ matches the area under the graph of $f$, that is, $f_{AVE} \cdot (b-a) = \int_a^b f(x)\,dx$. Dividing by $b-a$ we obtain definition (5.3).

**Interval splitting**

Suppose that the interval $[a,b]$ is split into two smaller intervals $[a,c]$ and $[c,b]$, where $c$ is some arbitrarily chosen number between $a$ and $b$. It is clear that

$$\int_a^b f(x)\,dx = \int_a^c f(x)\,dx + \int_c^b f(x)\,dx. \tag{5.4}$$

It would be convenient to be able to remove the restriction that $a \le c \le b$. For example, it should also be true that $\int_1^2 f(x)\,dx = \int_1^3 f(x)\,dx + \int_3^2 f(x)\,dx$. However, in the definition of the integral $\int_a^b f(x)\,dx$ we have assumed that $a \le b$. This is one of the reasons to adopt the following convention:

$$\int_b^a f(x)\,dx = -\int_a^b f(x)\,dx. \tag{5.5}$$

With this extension of the definition of the definite integral, the equality (5.4) holds for all values of $c$ (as long as all the integrals are well defined, of course).

## 5.2 The fundamental theorem of calculus

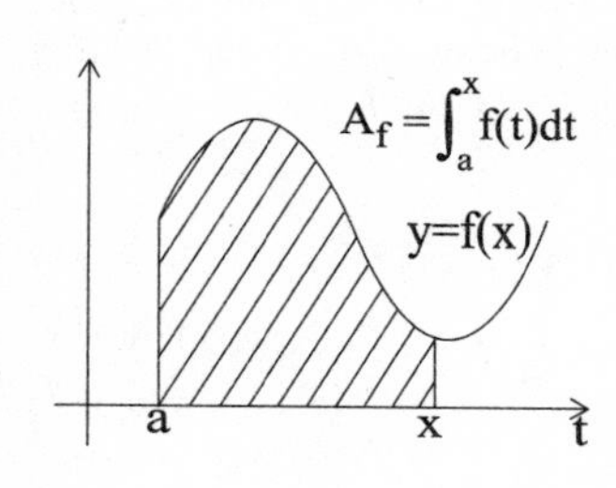


Let $f$ be a function and $a$ any point in its domain. For any input $x$, the **area function** $A_f$ is defined by the formula:

$$A_f(x) = \int_a^x f(t)\,dt.$$

In other words, $A_f(x)$ is the signed area defined by $f$ over $[a,x]$.

The definition of the area function $A_f(x)$ may seem intimidating at first, but it does make sense. It is also essential in the understanding of the fundamental theorem of calculus.

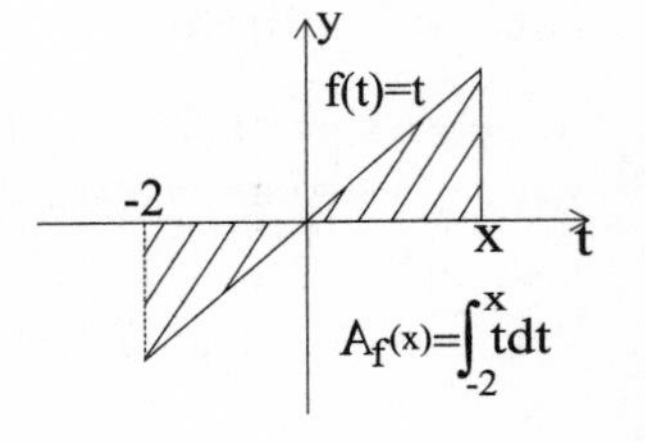


**Example 51** Let $g(t) = 5$ and $a = 3$. Define the area function $A_g(x) = \int_3^x 5dt$. Since the graph of $g$ is the horizontal line $y = 5$, the values of $A_g(x)$ are easily calculated: the area of a rectangle is base × height. We have $A_g(x) = (x-3) \cdot 5 = 5x - 15$. Let us notice that the derivative of the area function $\frac{d}{dx} A_g(x) = \frac{d}{dx}(5x - 15) = 5 = g(x)$.

**Example 52** Consider $f(t) = t$ and let $a = -2$. The area function $A_f(x) = \int_{-2}^x tdt$. For example, $A_f(4) = \int_{-2}^4 tdt = 6$, which can be calculated as the difference of the areas of two triangles. We leave it to the reader to check that $A_f(-2) = 0$, $A_f(0) = -2$ and $A_f(2) = 0$. In general, we have $A_f(x) = \frac{1}{2}x^2 - 2$. It is not hard to notice that, at least in this case, the derivative of the area function $\frac{d}{dx} A_f(x) = \frac{d}{dx}\left(\frac{1}{2}x^2 - 2\right) = x = f(x)$.

**Exercise 53** *Choose an arbitrary value of $a$, say $a = 1$. Calculate the value of the area function of a few simple functions, such as $f(t) = 3t + 5$, $g(t) = |t|$ or $k(t) = \begin{cases} 2 & for \quad t \leq 5, \\ t & for \quad t > 5. \end{cases}$*

Our goal is to understand how the area function is related to the underlying function $f$. Specifically, we are interested in the rate of change of $A_f(x)$, that is, in the derivative $\frac{d}{dx} A_f(x)$. We will assume that the function $f$ is continuous.

If $x$ is moved to the right by $h$, then the area under the graph of $f$ should increase by approximately the area of the rectangle with base $h$ and height $f(x)$. Since the function $f$ is continuous, the values of $f$ over the interval $[x, x+h]$ should all be fairly close to $f(x)$, the value at the left endpoint of the interval. We estimate the change in area $\Delta A_f(x) = A_f(x+h) - A_f(x) \approxeq f(x) \cdot h$. Hence, the rate of change of $A_f$ over the interval $[x, x+h]$

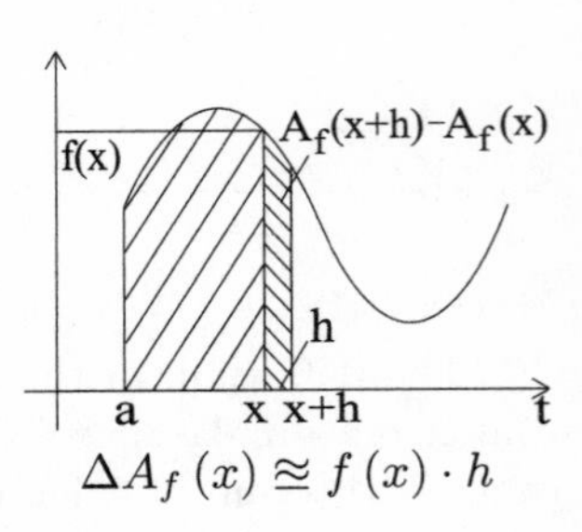

$\Delta A_f(x) \approxeq f(x) \cdot h$

$$\frac{\Delta A_f(x)}{h} = \frac{A_f(x+h) - A_f(x)}{h} \approxeq \frac{f(x) \cdot h}{h} = f(x),$$

and this approximation gets better and better as $h$ gets closer to 0. Taking the limit as $h \to 0$ we find the derivative $\frac{d}{dx} A_f(x) = \lim_{h \to 0} \frac{A_f(x+h) - A_f(x)}{h} = f(x)$. In a nutshell, the rate with which one can sweep the garage floor is proportional to the width of the broom. The simple argument above, while informal, contains the essence of the proof of the first version of the fundamental theorem of calculus:

**Theorem 54** *(FTC, version I) Let $f$ be continuous on $[a, b]$. For every $x \in [a, b]$ we have*

$$\frac{d}{dx} \int_a^x f(t)\, dt = f(x).$$

**Theorem 55** *(FTC, version II) Let $f$ be continuous on $[a, b]$, and let $F$ be any antiderivative of $f$. Then*

$$\int_a^b f(x)\, dx = F(b) - F(a).$$

The second version is used more frequently than the first, as it gives a method of calculating definite integrals. It also follows easily from the first. Suppose that $F$ is any antiderivative of $f$ (that is, some function with $F' = f$). Then $F$ can differ by at most a constant from the function $A_f$, which we know is an antiderivative of $f$ from the first version of the theorem, that is, $F(x) = A_f(x) + C$. We have

$$\begin{aligned} F(b) - F(a) &= (A_f(b) + C) - (A_f(a) + C) \\ &= A_f(b) - A_f(a) = \int_a^b f(x)\, dx - \int_a^a f(x)\, dx \\ &= \int_a^b f(x)\, dx - 0 = \int_a^b f(x)\, dx. \end{aligned}$$

In the opposite direction, assume that version II is true. From the definition of the area function, we have

$$\begin{aligned} A_f'(x) &= \frac{d}{dx} \int_a^x f(t)\, dt = \frac{d}{dx} (F(x) - F(a)) \\ &= F'(x) - 0 = f(x). \end{aligned}$$

The FTC gives justification to the commonly accepted notation for antiderivatives. The general antiderivative of a function $f$ is denoted by $\int f(x)\,dx$ and is called the **indefinite integral** of $f$. Once again, the indefinite integral $\int f(x)\,dx$ is *a family of functions* of the form $F(x)+C$, where $F'(x)=f(x)$, while the definite integral $\int_a^b f(x)\,dx$ is a *single number*: the area below the graph of $f$ over $[a,b]$.

**Example 56** Let us revisit the previous example and use the FTC to calculate $\int_{-2}^{4} tdt = F(4) - F(-2)$, where $F(x) = \frac{1}{2}x^2 + C$. We have $\int_{-2}^{4} tdt = \frac{1}{2}4^2 - \frac{1}{2}(-2)^2 = 8 - 2 = 6$. Typically we drop the constant $C$ when evaluating definite integrals, as it always gets cancelled in the process. A common notation for the above calculation is

$$\int_{-2}^{4} tdt = \frac{1}{2}t^2\Big|_{-2}^{4} = \frac{1}{2}4^2 - \frac{1}{2}(-2)^2 = 6.$$

**Example 57** To find the area under the graph of $f(x) = \frac{1}{x}$ over the interval $[1,10]$ we integrate: $\int_1^{10} \frac{1}{x}dx = (\ln x)|_1^{10} = \ln 10 - \ln 1 = \ln 10$. A side comment: one can define the natural logarithm without any reference to the magical number $e \approxeq 2.7183...$ Specifically, $\ln x = \int_1^x \frac{1}{t}dt$.

**Example 58** To find the area under the graph of $g(x) = \frac{1}{x^2+1}$ over $[0,1]$ we calculate: $\int_0^1 \frac{1}{x^2+1}dx = \arctan x|_0^1 = \frac{\pi}{4}$.

**Example 59** Let $F(x) = \int_0^x \frac{\sin t}{t}dt$. The task of finding an antiderivative of $\frac{\sin t}{t}$ is impossible, at least as a so-called elementary function. The reader can easily verify by differentiation that any simple pretender for an antiderivative of $\frac{\sin t}{t}$ is a usurper! However, we can still find the derivative $\frac{d}{dx}F(x) = \frac{\sin x}{x}$, by the FTC, version I.

The fundamental theorem of calculus makes it worth our while to find some methods for finding antiderivatives of functions. Compared to differentiation (finding derivatives) the process of integration (finding antiderivatives) is much harder. The good news is that there are many calculators and software packages capable of symbolic integration. This makes it somewhat less important to be able to do it all on paper. However, it is still useful to acquire some degree of proficiency in integration techniques.

## 5.3 Common integration mistakes

Let us begin with some good news. All the formulas for derivatives can be applied "in reverse." There is little point in memorizing various formulas of that sort, as long as we know how to differentiate.

**Example 60** We know that $\frac{d}{dx}\left(x^4\right) = 4x^3$. Therefore $\int 4x^3 dx = x^4 + C$. In order to integrate $x^\alpha$, we need to increase the exponent by one and multiply the result by the reciprocal of the new exponent:

$$\int x^\alpha dx = \frac{1}{\alpha+1} x^{\alpha+1} + C$$

if $\alpha \neq -1$. For $\alpha = -1$ we have of course $\int x^{-1} dx = \ln x + C$.

Integration is a linear operation:

$$\begin{aligned} \int f(x) + g(x)\, dx &= \int f(x)\, dx + \int g(x)\, dx, \\ \int k \cdot f(x)\, dx &= k \cdot \int f(x)\, dx. \end{aligned}$$

That was the good news. However, sooner or later we must come to terms with some really bad news: there are **NO RULES for how to integrate a product or a quotient of functions!** It may have taken a while to learn how to differentiate products and quotients of functions, but at least there were rules. Not so with integration. In particular,

$$\int f(x) \cdot g(x)\, dx \neq \int f(x)\, dx \cdot \int g(x)\, dx \tag{5.6}$$

and

$$\int \frac{f(x)}{g(x)} dx \neq \frac{\int f(x)\, dx}{\int g(x)\, dx}. \tag{5.7}$$

Needless to say, one cannot take square roots, or anything else except constants, outside the integral symbol, for example, $\int \sqrt{f(x)}dx \neq \sqrt{\int f(x)\, dx}$.

In general, it is much better for a calculus apprentice to say "I am sorry, but I am unable to find an antiderivative of this function," than to give an obviously wrong answer! Integration can be a complicated process and even the best of us make simple mistakes once in a while. However, one is well advised to check integrals by differentiation to avoid giving bad answers, which tend to frustrate those who read such erroneous solutions. The following examples are not intended to offend the reader's intelligence, but to illustrate the "anti-rules" (5.6), (5.7) and a few others. A more experienced reader can skip the following examples, however in the author's experience, the common mistakes below are routinely committed even by good students. We all need to learn to resist "obvious" answers.

**Example 61** Since $(e^x)' = e^x$, we have $\int e^x dx = e^x + C$. However, a common mistake is to attempt integration of functions of the form $e^{g(x)}$ in a naive way, for instance, $\int e^{5x} dx \neq e^{5x} + C$. By the chain rule $\frac{d}{dx}\left(e^{g(x)}\right) = e^{g(x)} \cdot g'(x) \neq e^{g(x)}$. In the case when $g(x) = kx$, we can easily fix the problem: $\int e^{kx} dx = \frac{1}{k}e^{kx} + C$, for any constant $k \neq 0$. Unfortunately, this simple fix works **only** when $g(x)$ is a linear function. The lesson is that in general

$$\int e^{f(x)} dx \neq e^{f(x)} + C \quad \text{and} \quad \int e^{f(x)} dx \neq \frac{e^{f(x)}}{f(x)} + C. \tag{5.8}$$

**Example 62** Continuing the previous example, $\int xe^{x^2} dx = \frac{1}{2}e^{x^2} + C$, which is easily verified by differentiation. Apparently, having the $x$ in front of $e^{x^2}$ is a blessing, not a curse. A common mistake beginners make is to think as follows: "I know how to integrate $x$. If I only knew how to integrate $e^{x^2}$, maybe I would make some progress in this integral." There are two problems with that reasoning. First, even if we knew how to integrate $e^{x^2}$, this would not directly help us integrate the product $xe^{x^2}$. A second problem is that it is not possible to write the integral of $e^{x^2}$ in terms of elementary functions. (There are tables of values of a closely related integral in every statistics book. They would not have been there if one could just find a nice antiderivative formula.)

**Example 63** In the same flavor, we know that $\int \frac{1}{x} dx = \ln x + C$ and that $\int \cos x dx = \sin x + C$. This does not help in finding the antiderivative of $\frac{\cos x}{x} = \frac{1}{x} \cdot \cos x$. It certainly is not $(\ln x) \cdot (\sin x)$, as $\frac{d}{dx}((\ln x) \cdot (\sin x)) = \frac{1}{x} \sin x + \cos x \ln x$. This illustrates (5.6).

**Example 64** Another common error is to misuse the fact that $(\ln x)' = \frac{1}{x}$. Yes, that means that $\int \frac{1}{x} dx = \ln x + C$ (OK, it is more general to write $\ln |x| + C$ instead, to allow for negative values of $x$). However,

$$\int \frac{1}{u(x)} dx \neq \ln u(x) + C. \tag{5.9}$$

Differentiating the right-hand side of (5.9) correctly, we get $\frac{d}{dx} \ln u(x) = \frac{1}{u(x)} \cdot u'(x) = \frac{u'(x)}{u(x)}$. In addition, we know that

$$\int \frac{1}{x^2+1} dx = \arctan x + C \neq \ln \left(1 + x^2\right) + C,$$

as opposed to

$$\int \frac{2x}{x^2+1} = \ln \left(x^2+1\right) + C.$$

**Example 65** To illustrate (5.7) let us consider $\frac{\cos x}{x}$. Some students attempt to integrate this quotient one piece at a time: $\int \frac{\cos x}{x} dx = \frac{\sin x}{\frac{1}{2}x^2} + C = \frac{2 \sin x}{x^2} + C$. This is of course a bad idea, since by the quotient rule the derivative $\frac{d}{dx}\left(\frac{2 \sin x}{x^2}\right) = \frac{2x \cos x - 4 \sin x}{x^3}$, which is not even close to $\frac{\cos x}{x}$.

There is a bit of good news: we do have a "poor man's product rule," officially called integration by parts (Section 6.2), but again, it does not quite solve the problem for a general product of functions. This and some other integration methods are discussed in Chapter 6.

# 5.4 Riemann sums and numerical approximation.

The purpose of this section is twofold: to define the definite integral properly, as a limit of the approximating Riemann sums, and to understand what a computer might be doing when it numerically calculates an integral. The general idea is to approximate the area of the region under the graph of a function by replacing it with a collection of rectangles (or trapezoids). We begin by partitioning the interval $[a,b]$ into $n$ subintervals by any $n+1$ points

$$a = x_0 < x_1 < \cdots < x_{n-1} < x_n = b.$$

Let $\Delta x_i = x_i - x_{i-1}$ and within each interval $[x_{i-1}, x_i]$ let us choose a sampling point $c_i$. The sum

$$S_n = f(c_1)\Delta x_1 + f(c_2)\Delta x_2 + \cdots + f(c_n)\Delta x_n = \sum_{i=1}^{n} f(c_i)\Delta x_i$$

is called the **Riemann sum** with $n$ subdivisions for $f$ on $[a,b]$.

We will consider a sequence of partitions $\mathcal{P}_n$ of the interval $[a,b]$ into subintervals. It is natural to expect that we get a better approximation of the actual integral when the value of $n$ is larger. It is not essential that all the subintervals have the same length, but to keep the partitions reasonable, we should require that the lengths of the subintervals $\Delta x_i$ are also getting smaller as $n \to \infty$.

We define the integral $\int_a^b f(x)\,dx$ as the number, if one exists, to which all Riemann sums $S_n$ tend, as $n$ tends to infinity and the widths of all subdivisions tend to zero. In symbols

$$\int_a^b f(x)\,dx = \lim_{n\to\infty} S_n = \lim_{n\to\infty} \sum_{i=1}^{n} f(c_i)\Delta x_i. \tag{5.10}$$

(This is a slight abuse of notation. We are not taking just one limit, but we allow all sequences of partitions, including the ones where the interval is divided unevenly. We also allow all possible choices of $c_i$, in the interval $[x_{i-1}, x_i]$. We define a norm $\|\mathcal{P}_n\|$ of a partition $P_n$ to be the length of the longest subinterval. In the definition of the integral we require that the norms of partitions tend to 0 with $n \to \infty$. For the integral to exist, the limit (5.10) should exist and be the same, no matter how the interval is cut or how the $c_i$'s are chosen. These details are typically omitted in calculus courses and are left for more advanced classes.)

While this approach to the concept of a definite integral makes sense, the definition does not rely on any intuition about area. Instead of using an intuitive concept of the area of a region to define the integral, we define the integral as a limit of a sequence of Riemann sums. It gets better: we can now define certain areas as integrals. This not only makes the definition precise but also allows us to compute the integral numerically, which is necessary whenever the function is too difficult to integrate symbolically. **Numerical integration** is also needed when the function is nothing but a collection of data in a table, which is typical for many real life applications.

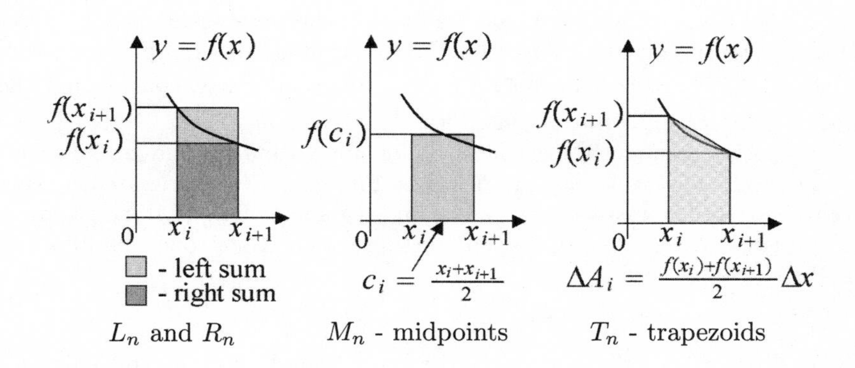


$L_n$ and $R_n$ $M_n$ - midpoints $T_n$ - trapezoids

For certain systematic ways of choosing the point $c_i$ in each subinterval, we define the so-called right sum ($R_n$), left sum ($L_n$) and midpoint sum ($M_n$). These names are self-explanatory. For numerical approximations we also frequently use the trapezoid sum ($T_n$), in which trapezoids are used in place of the rectangles. It is not hard to verify that $T_n = \frac{1}{2}(L_n + R_n)$.

**Example 66** Consider a function $f(x) = x^2$ on the interval $[2, 5]$. The exact value of the integral $\int_2^5 x^2 dx = \frac{1}{3}x^3\Big|_2^5 = \frac{1}{3}5^3 - \frac{1}{3}2^3 = 39$. Let us divide the interval into three subintervals of the same length. Depending on the choice of points $c_i$, we obtain the following approximations:

$$\begin{aligned}
L_3 &= [f(2) + f(3) + f(4)] \cdot \Delta x = \left[2^2 + 3^2 + 4^2\right] \cdot 1 = 29, \\
R_3 &= [f(3) + f(4) + f(5)] \cdot \Delta x = \left[3^2 + 4^2 + 5^2\right] \cdot 1 = 50, \\
M_3 &= [f(2.5) + f(3.5) + f(4.5)] \cdot \Delta x = 2.5^2 + 3.5^2 + 4.5^2 = 38, \\
T_3 &= \frac{L_3 + R_3}{2} = \frac{29 + 50}{2} = \frac{79}{2} = 39.5.
\end{aligned}$$

The approximate values obtained in the previous example by the methods of trapezoids and midpoints are much better than the other two. This is not a big surprise: the left and right sum approximations are a little crude as they replace the values of the function $f$ on every subinterval by the value at the end of the interval. Choosing the midpoint is more sensible. The trapezoid sum $T_n$ results from replacing the graph of the function on each subinterval with the straight line segment connecting the points $(x_k, f(x_k))$ and $(x_{k+1}, f(x_{k+1}))$, which should also lead to better results. It is also clear from geometrical considerations that if a function is, for example, increasing, then $L_n < \int_a^b f(x)\,dx < R_n$, for all $n$. If a function is, say, concave up, then $M_n < \int_a^b f(x)\,dx < T_n$ (see Problem 8).

Yet another frequently used method for approximating integrals is Simpson's method. This time, the graph of the function is approximated by small pieces of parabolas, instead of straight line segments. This is done two intervals at a time. After some (not so interesting) calculations, one can obtain the formula

$$S_n = \frac{2}{3}T_{2n} + \frac{1}{3}M_{2n}.$$

Any estimate is only as good as the error bound. For the convenience of the reader, the error bounds are provided below:

$$|I - L_n| \le \tfrac{K_1(b-a)^2}{2n}, \qquad |I - R_n| \le \tfrac{K_1(b-a)^2}{2n},$$
$$|I - T_n| \le \tfrac{K_2(b-a)^3}{12n^2}, \qquad |I - M_n| \le \tfrac{K_2(b-a)^3}{24n^2},$$
$$|I - S_n| \le \tfrac{K_4(b-a)^5}{180n^4},$$

where the constants $K$ are bounds on the values of appropriate derivatives on the interval of integration, that is, $|f'(x)| \le K_1$, $|f''(x)| \le K_2$ and $\left|f^{(4)}(x)\right| \le K_4$ for $x \in [a,b]$.

The formulas for the error bounds are not very difficult to derive, but we will not do that here, for the sake of brevity. Informally, the bounds on the values of derivatives give constraints on how badly the actual area can stray from the area under the approximating function. In the method of left and right rectangles, the function was approximated by a constant on each subinterval and the bound on $f'$ tells us how far from that constant the values of $f$ could get. In the case of trapezoids, we approximate the function by a straight line on each subinterval. The second derivative controls how far from that straight line can we stray, and so on.

## 5.5 Improper integrals

All definite integrals considered so far were on finite intervals and the functions were bounded. This restriction can be removed to allow for a function to have a vertical asymptote or for the interval of integration to be infinite.

The first type of improper integral is when the interval of integration has infinite length, say $(a, \infty)$. Rather than try to modify the definition of the integral to allow for the interval to be infinite, we define

$$\int_a^\infty f(x)\,dx = \lim_{t\to\infty} \int_a^b f(x)\,dx. \tag{5.11}$$

In other words, we approximate the improper integral $\int_a^\infty f(x)\,dx$ by the proper ones over large intervals $[a, t]$. When the limit is finite, we say that the integral **converges;** when it is infinite, we say that it **diverges to** $\infty$ (or to $-\infty$). Finally, if the limit (5.11) does not exist, we simply say that the integral **diverges** (we do not say to what, as there is no limit).

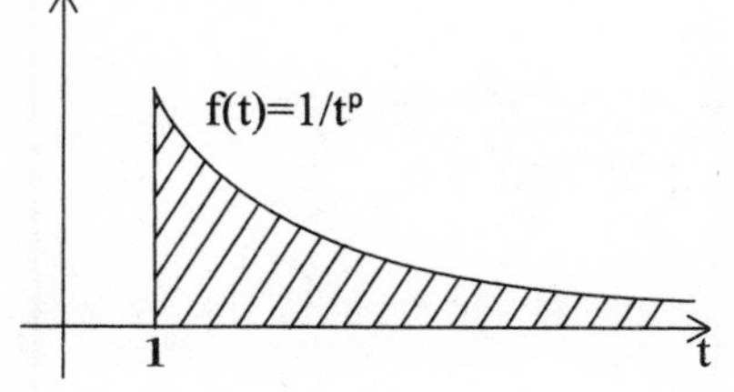


**Example 67** Let us find the area below the graph of $f(x) = \frac{1}{x}$ over the interval $[1, \infty)$. We have $\int_1^\infty \frac{1}{x} dx = \lim_{t\to\infty} \int_1^t \frac{1}{x} dx = \lim_{t\to\infty} (\ln t) = \infty$. The area is infinite.

**Example 68** The result is different for $f(x) = \frac{1}{x^2}$. We have $\int_1^\infty \frac{1}{x^2} dx = \lim_{t\to\infty} \int_1^t \frac{1}{x^2} dx = \lim_{t\to\infty} \left(-\frac{1}{x}\right)\big|_1^t = \lim_{t\to\infty} \left(-\frac{1}{t} + 1\right) = 1$.

**Example 69** Generalizing the previous example, consider any value of $p \neq 1$. We have

$$\begin{aligned}\int_1^\infty \frac{1}{x^p}dx &= \lim_{t\to\infty}\int_1^t x^{-p}dx = \lim_{t\to\infty}\left(\frac{1}{-p+1}x^{-p+1}\Big|_1^t\right) \\ &= \frac{1}{-p+1}\lim_{t\to\infty}\left(\frac{1}{x^{p-1}}\Big|_1^t\right) = \frac{1}{-p+1}\lim_{t\to\infty}\left(\frac{1}{t^{p-1}}-1\right).\end{aligned}$$

It is easy to verify that the limit is $\infty$ if $p < 1$ and that it is finite when $p > 1$. Combining this with Example 67 we obtain

$$\int_1^\infty \frac{1}{x^p}dx = \begin{cases} \frac{1}{p-1} & \text{if } p > 1, \\ \infty & \text{if } p \leq 1. \end{cases} \tag{5.12}$$

Example 69 is worth contemplating for a while. The graphs of functions of the form $f(x) = \frac{1}{x^p}$ all have similar shape for $p > 0$. However, the area under the graph of $f$ could be finite or infinite, depending on the value of the exponent $p$. In both cases $f$ asymptotically approaches 0 as $x \to \infty$, but for the area to be finite, this needs to happen fast enough.

Another type of improper integral is needed when a function has a vertical asymptote, say at $x = a$. In this case, for a function on the interval $(a, b)$, we approximate the improper integral $\int_a^b f(x)\,dx$ by integrals $\int_t^b f(x)\,dx$. Formally, $\int_a^b f(x)\,dx = \lim_{t\to\infty}\int_t^b f(x)\,dx$.

**Example 70** Let $f(x) = \frac{1}{\sqrt{x}}$. Since there is a vertical asymptote at $x = 0$, we define $\int_0^1 \frac{1}{\sqrt{x}}dx = \lim_{t\to 0^+}\int_t^1 \frac{1}{\sqrt{x}}dx = \lim_{t\to 0^+}\left(2\sqrt{x}\big|_t^1\right) = 2$.

**Exercise 71** *As in Example 69, determine for what values of p, the integral $\int_1^\infty \frac{1}{x^p}dx$ is a finite number. (Ans.: for $p < 1$.)*

Finally, when the integral is improper for several reasons, the interval of integration needs to be split into smaller ones, on which the integral is "singly improper." For example, the integral $\int_0^\infty \frac{1}{x^2}dx$ is defined as the sum $\int_0^1 \frac{1}{x^2}dx + \int_1^\infty \frac{1}{x^2}dx$. The second integral converges to 1, but since the first one diverges to $\infty$, the original integral also diverges to $\infty$. (There is nothing special about the choice of number 1 as the splitting point; we could have chosen any other positive number instead.)

**Exercise 72** *Consider $\int_{-\infty}^{\infty} \frac{1}{x} dx$. This integral is improper on both ends of the interval. In this case, the interval of integration needs to be split by an arbitrarily chosen point, say $x = 1$. We write $\int_{-\infty}^{\infty} \frac{1}{x} dx = \int_{-\infty}^{1} \frac{1}{x} dx + \int_{1}^{\infty} \frac{1}{x} dx$, which means that in order for the integral $\int_{-\infty}^{\infty} \frac{1}{x} dx$ to exist both integrals over smaller intervals have to exist. However, we already know that $\int_{1}^{\infty} \frac{1}{x} dx = \infty$, so the integral $\int_{-\infty}^{\infty} \frac{1}{x} dx$ diverges (does not exist). One needs to be careful to avoid the temptation of writing $\int_{-\infty}^{\infty} f(x) dx = \lim_{t\to\infty} \int_{-t}^{t} f(x) dx$ (see problem 6).*

### 5.5.1 Comparison test

Sometimes we need to determine if an improper integral converges, even though we may be unable to calculate its exact value. The main idea is to replace a more complicated integral with a simpler one whose convergence we can determine. However, one has to be careful not to jump to the wrong conclusions. To show convergence of a given integral, we should look for a friendly convergent integral greater than the one in question. In the opposite direction, to show divergence of a given integral, we should find a smaller integral which diverges.

**Example 73** To see that the integral $\int_{3}^{\infty} \frac{1}{x^2+5} dx$ converges, we just need to notice that $\frac{1}{x^2+5} < \frac{1}{x^2}$. By (5.12) we know that $\int_{3}^{\infty} \frac{1}{x^2+5} dx < \int_{3}^{\infty} \frac{1}{x^2} dx < \infty$.

**Example 74** We need to be a little more careful with $\int_{2}^{\infty} \frac{1}{x^2-5} dx$, as unfortunately $\frac{1}{x^2-5} > \frac{1}{x^2}$ and it really does not help us to know that $\int_{2}^{\infty} \frac{1}{x^2-5} dx > \int_{3}^{\infty} \frac{1}{x^2} dx$. Yes, the second integral is finite but ours is larger. The inequality is true, but useless! However, at least for sufficiently large $x$, we have $\frac{1}{x^2-5} < \frac{1}{x^2-\frac{1}{2}x^2} = \frac{2}{x^2}$. Now we can conclude that $\int_{2}^{\infty} \frac{1}{x^2-5} dx$ converges.

Using the comparison test sometimes takes some creativity. We have some freedom in what to compare our functions with. The functions which are frequently used for that purpose are $\frac{1}{x^p}$ and $e^{-x}$, or some of their variations (see problem 7). We will come back to this issue in Section 8.3.

## 5.6 Problems

1. Integrate $\int_0^2 \sqrt{4-x^2}dx$.

2. Explain why the integral $\int_{-3}^{3} \sin\left(x^3\right) dx = 0$.

3. Find the derivative of the functions:

   (a) $F(x) = \int_x^5 \frac{e^t}{t} dt$,

   (b) $G(x) = \int_1^{x^3} \frac{\sin t}{t} dt$,

   (c) $H(x) = \int_{2x}^{x^2} \sin\left(t^2\right) dt$.

4. Give your own examples to illustrate the formulas (5.6), (5.7), (5.8) and (5.9).

5. Show by direct computation of the limit that the integral $\int_0^\infty e^{-2x} < \infty$.

6. Explain why $\lim_{t\to\infty} \int_{-t}^{t} xdx = 0$, but the integral $\int_{-\infty}^{\infty} xdx$ diverges.

7. Use the comparison test to show that the integrals converge:

   (a) $\int_1^\infty \frac{1}{x^2-x} dx$

   (b) $\int_\pi^\infty \frac{\sin x}{x^2} dx$

   (c) $\int_0^\infty \frac{e^{-x}}{\sqrt{x}} dx$ (Hint: the interval needs to be split).

8. Make a convincing argument that if a function is concave up on $[a, b]$, then the method of trapezoids overestimates the value of the integral $\int_a^b f(x)\, dx$, and the midpoint approximation underestimates it. Use the fact that the tangent line lies below, and the secant line lies above, the graph of the function, due to concavity. Notice that the area of the "midpoint rectangle" is the same as the area of the trapezoid below the appropriate tangent line and above the interval $[a, b]$.

# Chapter 6

# Methods of Integration

As we have seen in Section 5.3, integrating can be quite tricky. There is no general method to integrate a product or a quotient of functions. In spite of this disappointing fact, in this chapter we look at several possible approaches to finding antiderivatives. Still, it is common to run into a function which cannot be integrated using any known method. Worse still, there are many innocent-looking functions whose integrals cannot be expressed in closed form (i.e., using the so-called elementary functions and a finite sequence of operations such as addition, subtraction, multiplication, division, raising to powers, or composing with other functions). For instance, $\int \frac{\sin x}{x} dx$ and $\int e^{-x^2} dx$ cannot be expressed in closed form. On the bright side, there are many functions, which can be integrated and a few integrating "tricks" will now be described. The importance of the integration techniques is somewhat diminished due to the availability of calculators and computer algebra systems capable of performing symbolic integration. However, it is useful to be proficient in at least some of the integration methods.

## 6.1 Method of substitution

A composition of functions $g(u(x))$ can be differentiated by the chain rule: $\frac{d}{dx}[g(u(x))] = g(u(x)) \cdot u'(x)$. The method of substitution is the chain rule in reverse: if $f(x) = g(u(x)) \cdot u^{'}(x)$, then

$$\int f(x)\,dx = \int g(u)\,du, \tag{6.1}$$

and for definite integrals

$$\int_a^b f(x)\,dx = \int_{u(a)}^{u(b)} g(u)\,du. \tag{6.2}$$

It is important to realize that the method of substitution allows us to integrate the expression $g(u(x)) \cdot u^{'}(x)$ and not just the composition of the functions $g(u(x))$. The factor $u^{'}(x)$ is essential. The general formulas look quite intimidating and are best explained by examples.

**Example 75** Consider $F(x) = \left(x^3+5\right)^7$. Then $F'(x) = 7\left(x^3+5\right)^6 \cdot 3x^2$. In the language of integrals, we have $\int 7\left(x^3+5\right)^6 \cdot 3x^2dx = \left(x^3+5\right)^7 + C$. In this example $f(x) = 7\left(x^3+5\right)^6 \cdot 3x^2 = 7u^6 \cdot u'$, where $u = x^3+5$ and $u' = 3x^2$. The factor $3x^2$ is essential.

**Example 76** To find $\int 3x^2e^{x^3}dx$ we substitute $u = x^3$. Then $\frac{du}{dx} = 3x^2$ and a common notation is to write $du = 3x^2dx$. We have

$$\int 3x^2e^{x^3}dx = \int e^{x^3}3x^2dx = \int e^u du = e^u + C = e^{x^3} + C.$$

While the notation looks a little mysterious at first, we can easily check that the answer is correct: $\frac{d}{dx}\left(e^{x^3}\right) = e^{x^3} \cdot 3x^2$.

**Example 77** This time let us consider a definite integral $\int_1^e \frac{(\ln x)^2}{x}dx$. The substitution $u = \ln x$ is a natural choice, as $\frac{du}{dx} = \frac{1}{x}$. We have

$$\int \frac{(\ln x)^2}{x}dx = \int u^2du = \frac{1}{3}u^3 + C = \frac{1}{3}(\ln x)^3 + C,$$

and for the definite integral we have

$$\int_1^e \frac{(\ln x)^2}{x} dx = \frac{1}{3}(\ln x)^3\Big|_1^e = \frac{1}{3}(\ln e)^3 - \frac{1}{3}(\ln 1)^3 = \frac{1}{3}\cdot 1 - 0 = \frac{1}{3}.$$

A better way to perform the substitution for a definite integral is to change the limits of integration and never have to come back to the original variable:

$$\int_1^e \frac{(\ln x)^2}{x} dx = \int_0^1 u^2 du = \frac{1}{3}u^3\Big|_0^1 = \frac{1}{3},$$

where the new limits of integration are obtained by plugging the original limits for $x$ into the new variable $u = \ln x$. Changing the limits makes the solution shorter and more elegant. (Doesn't it feel good to make a clean break from your $x$? )

A formal proof of the substitution rule is not very enlightening. It basically restates the chain rule in terms of antiderivatives. Geometrically speaking, the substitution $u = u(x)$ means "stretching" or "shrinking" the $x$-axis. The region under the graph of $y = f(x)$ remains essentially the same in the new variables $y = g(u)$; however, the horizontal axis $x$ is "stretched" by a factor of $\frac{du}{dx}$, which is why it is needed under the integral.

### 6.1.1 What to substitute?

After we understand the method of substitution in principle, the main question is what should we substitute? The answer depends on the integral. The right choice for the substitution is not always obvious and sometimes it takes some ingenuity to find the right substitution. There are some things we need to watch for. Typically, we look for functions whose derivative occurs in the function we are trying to integrate. If we substitute $u(x)$, it is essential that the expression for the derivative $\frac{du}{dx}$ shows up as a factor somewhere under the integral. The substitution is not complete until everything is expressed in terms of the new variable. There should be no "leftovers" of the old variable. We must remember that $dx$ cannot be simply replaced by $du$, but $u'dx$ can be replaced by $du$. In Example 76 it was critical that the expression for $\frac{du}{dx} = 3x^2$ was in the original integral. Without $3x^2$, the integral $\int e^{x^3} dx$ is impossible to handle. And finally, in integration, as in everything else we do in mathematics, the rules of algebra must be observed religiously.

Some fractions can also be integrated quite easily using substitution. To integrate a fraction in which the numerator happens to be the derivative of the denominator, we substitute the denominator and get

$$\int \frac{u'(x)}{u(x)} dx = \int \frac{1}{u} du = \ln u + C \tag{6.3}$$

(but $\int \frac{1}{u(x)} dx \neq \ln u + C$, as shown in Example 64 in Section 5.3).

**Example 78** How to integrate $\int \frac{1+2x}{1+x^2} dx$? It would seem natural to substitute $u = 1 + x^2$, however $du = 2x dx$. What about the 1? We cannot just ignore it (although some students try). The trick is to split the integral:

$$\int \frac{1+2x}{1+x^2} dx = \int \frac{1}{1+x^2} dx + \int \frac{2x}{1+x^2} dx.$$

The first integral is just $\arctan x$ and the second one is done by substitution $u = 1 + x^2$:

$$\int \frac{2x}{1+x^2} dx = \int \frac{du}{u} = \ln u + C = \ln\left(1 + x^2\right) + C,$$

as we have already seen in example 64 before officially calling it a substitution.

**Example 79** To integrate $\int_{-4}^{5} x\sqrt{x+4} dx$, we substitute $u = x + 4$. The new variable $u$ changes at the same rate as the original: $du = dx$. However, we can now put $u$ under the radical and replace $x$ with $u-4$, which gives $\int_0^9 x\sqrt{x+4} dx = \int_0^9 (u-4)\sqrt{u} du = \int_0^9 \left(u^{3/2} - 4u^{1/2}\right) du = \left[\frac{2}{5}u^{5/2} - \frac{8}{3}u^{3/2}\right]\Big|_0^9 = \frac{2}{5}9^{5/2} - \frac{8}{3}9^{3/2} = \frac{2}{5}3^5 - \frac{8}{3}3^3 = 25.2$. The point of this substitution was that we could open parentheses after changing the variables, which was not possible when $x + 4$ was under the radical.

It takes many more examples than we can fit into this short book to master the method of substitution. A few exercises are provided at the end of this chapter, but the reader is encouraged to work on more examples, which can be found in every calculus textbook.

## 6.2 Integration by parts: the "poor man's product rule"

We learned in Section 3.2.1 that for any two differentiable functions $u(x)$ and $v(x)$ we have

$$(u(x) \cdot v(x))' = u'(x) \cdot v(x) + u(x) \cdot v'(x),$$

or $(u \cdot v)' = u' \cdot v + u \cdot v'$ for short. Integrating both sides we obtain

$$u(x) \cdot v(x) = \int u'(x) \cdot v(x)\, dx + \int u(x) \cdot v'(x)\, dx.$$

After subtracting one of the integrals from both sides we get

$$\int u(x) \cdot v'(x)\, dx = u(x) \cdot v(x) - \int u'(x) \cdot v(x)\, dx. \tag{6.4}$$

Typically, we write this in an equivalent, shorter form:

$$\int u dv = uv - \int v du, \tag{6.5}$$

where $dv$ stands for $v'(x)\, dx$ and $du$ for $u'(x)\, dx$. This makes sense, as $v'(x)\, dx = \frac{dv}{dx} dx$, so if the notation was designed sensibly, then we should be able to "cancel" the $dx$. The point of integration by parts is to find two functions $u$ and $v$, such that, after applying formula (6.5), the new integral is easier to find than the original. The trick is to choose the functions wisely (think of Indiana Jones choosing the chalice).

Knowing that there are such elaborate techniques for integrating products, one should never attempt to integrate products one piece a time: $\int fg dx = \int f dx * \int g dx$. It makes no sense, as is easily seen by considering just about any pair of functions $f$ and $g$ (see Section 5.3).

**Example 80** To integrate $\int xe^x dx$, introduce $u = x$. This leaves us little choice for $v$, as we need $dv = e^x dx$, which means that $v' = e^x$. Then $du = u'dx = 1 \cdot dx = dx$ and $v = e^x$. We have

$$\int xe^x dx = xe^x - \int e^x dx = xe^x - e^x + C = e^x (x - 1) + C.$$

**Example 81** In a similar fashion, to integrate $\int x \cos x dx$, we take $u = x$ and $dv = \cos x dx$. This leads to $du = dx$ and $v = \sin x$. We obtain

$$\int x \cos x dx = x \sin x - \int \sin x dx = x \sin x + \cos x + C.$$

**Example 82** Integrating $\int \ln x dx$ takes a little more ingenuity: there is no apparent product of two functions. However, we can think of $\ln x$ as $(\ln x) \cdot 1$. We take $u = \ln x$ and $dv = 1 \cdot dx = dx$. This gives $du = \frac{1}{x} dx$ and $v = x$. The expression $vdu$ simplifies very nicely as $x \cdot \frac{1}{x} dx = 1dx = dx$. Integration by parts gives us

$$\int \ln x dx = x \ln x - \int 1dx = x \ln x - x + C = x (\ln x - 1) + C.$$

Another possible strategy is to come back to the original integral, with a coefficient different from 1, usually after integrating by parts twice. Next, we solve the equation for the unknown integral. The following example illustrates this technique.

**Example 83** Let $I = \int e^{2x} \sin x dx$. We integrate by parts with $u = e^{2x}$ and $dv = \sin x dx$. This gives $du = 2e^{2x}$ and $v = -\cos x$, so

$$I = -e^{2x} \cos x - \int -2e^{2x} \cos x dx = -e^{2x} \cos x + 2 \int e^{2x} \cos x dx. \quad (6.6)$$

We now integrate the second integral by parts again, this time using the same $u = e^{2x}$ but a new $dv = \cos x dx$. This time $du$ is still $2e^{2x}$ and $v = \sin x$. We get

$$\int e^{2x} \cos x dx = e^{2x} \sin x - \int 2e^{2x} \sin x dx, \quad (6.7)$$

and substituting (6.7) into (6.6) we have

$$\begin{aligned} I &= -e^{2x} \cos x + 2 \left( e^{2x} \sin x - \int 2e^{2x} \sin x dx \right) \\ &= -e^{2x} \cos x + 2e^{2x} \sin x - 4 \int e^{2x} \sin x dx \\ &= -e^{2x} \cos x + 2e^{2x} \sin x - 4 \cdot I. \end{aligned}$$

Even though it seems that we have made no progress, we can treat this as an equation with an unknown $I$. This gives

$$I + 4I = -e^{2x}\cos x + 2e^{2x}\sin x$$

and dividing both sides by 5 we get the answer

$$I = -\frac{1}{5}e^{2x}\cos x + \frac{2}{5}e^{2x}\sin x + C.$$

**Example 84** On a slightly more advanced level, a similar technique can be used to obtain some of the "reduction" formulas. Let $n \geq 2$ be an integer and suppose that we are trying to integrate $I_n = \int \sin^n(x)\,dx$. We will think of $\sin^n(x)$ as a product,

$$\sin^n(x) = \sin^2(x)\cdot\sin^{n-2}(x) = \left(1 - \cos^2(x)\right)\cdot\sin^{n-2}(x).$$

The integral $I_n$ can be written as

$$\begin{aligned} I_n &= \int \sin^{n-2}(x)\,dx - \int \cos^2(x)\cdot\sin^{n-2}(x)\,dx \\ &= I_{n-2} - \int \cos^2(x)\cdot\sin^{n-2}(x)\,dx. \end{aligned}$$

Integration by parts is now used for the second integral, with $u = \cos x$ and $dv = \cos x \cdot \sin^{n-2}(x)\,dx$. It is not hard to see that $v = \frac{1}{n-1}\sin^{n-1}(x)$, and of course $du = -\sin x dx$. We get $vdu = -\frac{1}{n-1}\sin^{n-1}(x)\sin x dx = -\frac{1}{n-1}\sin^n(x)\,dx$ and

$$\int \cos^2(x)\cdot\sin^{n-2}(x)\,dx = \frac{1}{n-1}\cos x \sin^{n-1}(x) + \frac{1}{n-1}I_n;$$

therefore

$$I_n = I_{n-2} - \frac{1}{n-1}\cos x \sin^{n-1}(x) - \frac{1}{n-1}I_n.$$

Again, treating this as an equation with an unknown $I_n$ and solving for $I_n$ we get

$$I_n + \frac{1}{n-1}I_n = I_{n-2} - \frac{1}{n-1}\cos x \sin^{n-1}(x)$$

or

$$\frac{n}{n-1} I_n = -\frac{1}{n-1} \cos x \sin^{n-1}(x) + I_{n-2};$$

thus

$$I_n = -\frac{1}{n} \cos x \sin^{n-1}(x) + \frac{n-1}{n} I_{n-2},$$

in other words

$$\int \sin^n(x)\, dx = -\frac{1}{n} \cos x \sin^{n-1}(x) + \frac{n-1}{n} \int \sin^{n-2}(x)\, dx.$$

**Exercise 85** *Use integration by parts to establish the following reduction formulas:*

$$\int \cos^n x dx = \frac{\cos^{n-1} x \sin x}{n} + \frac{n-1}{n} \int \cos^{n-2} x dx, \tag{6.8}$$

$$\int \tan^n x dx = \frac{\tan^{n-1} x}{n-1} - \int \tan^{n-2} x dx,$$

$$\int \sec^n x dx = \frac{\sec^{n-2} x \tan x}{n-1} + \frac{n-2}{n-1} \int \sec^{n-2} x dx.$$

To finish this section, let us notice that in integration by parts we do not strive to get the most general possible function $v$, that is, we usually do not worry about adding the constant $C$, until the very last step. After all, we are not looking to find all possible functions $u$ and $v$, that might be used, but just one pair that works. The final answer for **indefinite** integrals should always contain an arbitrary constant $C$. This of course should not be done with a definite integral, for which the answer is a single number. (Adding a constant is just as meaningful as saying my car is worth $10,000, plus an arbitrary constant C.)

## 6.3 Integrating rational functions, partial fractions

It would be nice to have a complete class of functions we can integrate. So far, we can integrate all polynomials, that is, functions of the form $P(x) =$

$a_nx^n + a_{n-1}x^{n-1} + \cdots + a_1x + a_0$. A natural next step is to consider rational functions, which are quotients of polynomials.

A reciprocal of a linear function is easy to integrate, since $\frac{d}{dx}(\ln|ax+b|) = \frac{a}{ax+b}$. Therefore, after adjusting by the coefficient $\frac{1}{a}$ we obtain

$$\int \frac{1}{ax+b}dx = \frac{1}{a}\ln|ax+b| + C. \tag{6.9}$$

**Example 86** The integral $\int \frac{5}{3x+7}dx = \frac{5}{3}\ln|3x+7| + C$.

Integration of a rational function with a quadratic denominator is more complicated. In particular, $\int \frac{1}{x^2+1}dx = \arctan x + C$ and $\int \frac{2x}{x^2+1}dx = \ln(x^2+1) + C$, which already indicates that something interesting is going on. Let us consider a few examples.

**Example 87** How to integrate a function $f(x) = \frac{4x+5}{x^2+2x+2}$? The trick is to complete the square in the denominator: $x^2+2x+2 = (x+1)^2+1$. After making a substitution $u = x+1$, we get $du = dx$ and $4x+5 = 4u+1$, so

$$\begin{aligned}\int \frac{4x+5}{x^2+2x+2}dx &= \int \frac{4u+1}{u^2+1}du = \int \frac{4u}{u^2+1}du + \int \frac{1}{u^2+1}du \\ &= 2\ln(u^2+1) + \arctan u + C \\ &= 2\ln(x^2+2x+2) + \arctan(x+1) + C.\end{aligned}$$

**Example 88** Consider $g(x) = \frac{4x+5}{x^2+2x+3}$. This time $x^2+2x+3 = (x+1)^2+2$. As in the previous example, we substitute $u = x+1$ and obtain

$$\int \frac{4x+5}{x^2+2x+3}dx = \int \frac{4u+1}{u^2+2}du = \int \frac{4u}{u^2+2}du + \int \frac{1}{u^2+2}du.$$

The first integral $\int \frac{4u}{u^2+2}du = 2\ln(u^2+1) + C$, but the second one requires a simple trick. We know how to handle $\frac{1}{u^2+1}$. We need somehow to change 2 into 1 in $\frac{1}{u^2+2}$. Since

$$\frac{1}{u^2+2} = \frac{1}{2\left(\frac{u^2}{2}+1\right)} = \frac{1}{2}\cdot\frac{1}{\left(\frac{u}{\sqrt{2}}\right)^2+1},$$

another substitution is needed: $t = \frac{u}{\sqrt{2}}$ gives $\frac{1}{u^2+2} = \frac{1}{2} \cdot \frac{1}{t^2+1}$ and $dt = \frac{1}{\sqrt{2}} du$. Hence

$$\begin{aligned}
\int \frac{1}{u^2+2} du &= \int \frac{1}{2} \cdot \frac{1}{t^2+1} \sqrt{2} dt \\
&= \frac{\sqrt{2}}{2} \arctan(t) + C = \frac{1}{\sqrt{2}} \arctan\left(\frac{u}{\sqrt{2}}\right) + C \\
&= \frac{1}{\sqrt{2}} \arctan\left(\frac{x+1}{\sqrt{2}}\right) + C.
\end{aligned}$$

The denominators in examples 87 and 88 do not have any real roots - they are irreducible. In other words, after completing the squares, the denominator looked like $u^2 + k$, with $k > 0$. The situation is entirely different when the quadratic in denominator can be factored, as illustrated by the next example.

**Example 89** In order to find $\int \frac{2}{x^2-1} dx$, let us notice that

$$\begin{aligned}
\frac{1}{x-1} - \frac{1}{x+1} &= \frac{x+1}{(x+1)(x-1)} - \frac{x-1}{(x+1)(x-1)} \\
&= \frac{x+1-x+1}{(x+1)(x-1)} = \frac{2}{x^2-1}.
\end{aligned}$$

(Never mind for the moment how we came up with that.) Therefore

$$\begin{aligned}
\int \frac{2}{1-x^2} dx &= \int \frac{1}{x-1} dx - \int \frac{1}{x+1} dx \\
&= \ln|x-1| - \ln|x+1| + C = \ln\left|\frac{x-1}{x+1}\right| + C.
\end{aligned}$$

In Example 89 a rational function was written as a sum of two simpler fractions which we already knew how to integrate. This is a method by which we can integrate any rational function. What we need is a family of relatively simple rational functions which can be used as "building blocks." The functions in this family should be as simple as possible, yet complicated enough to allow for any rational function to be represented as a sum of those simpler "blocks."

We say that a rational function is **proper** when the degree of the numerator is smaller than the degree of the denominator.

**Example 90** The function $k(x) = \frac{x^2+1}{x^2-1}$ is not a proper rational function as the degrees of the numerator and denominator are both 2. However, since $x^2+1 = x^2-1+2$, we have

$$\begin{aligned} k(x) &= \frac{x^2+1}{x^2-1} = \frac{x^2-1+2}{x^2-1} = \\ &= \frac{x^2-1}{x^2-1} + \frac{2}{x^2-1} = 1 + \frac{2}{x^2-1}. \end{aligned}$$

By Example 89 we have

$$\int \frac{x^2+1}{x^2-1} dx = \int \left(1 + \frac{2}{x^2-1}\right) dx = x + \ln\left|\frac{x-1}{x+1}\right| + C.$$

In general, any improper rational function $\frac{p(x)}{q(x)}$ can be written as a sum of a polynomial and a proper rational function. Specifically, when the degree of the numerator $p(x)$ is not smaller than the degree of the denominator $q(x)$, we perform the division

$$\frac{p(x)}{q(x)} = quotient(x) + \frac{remainder(x)}{q(x)}.$$

To keep this book short, we do not discuss the process of long or synthetic division here. This topic can be found in most algebra textbooks. However, in many problems one can take some shortcuts in the division process.

**Example 91** Let $f(x) = \frac{x^3}{x^2+1}$. We would like to rewrite $x^3$ as $(x^2+1)\cdot \_\_ + \_\_$. We have $(x^2+1)\cdot x = x^3 + x$, which gives us the desired $x^3$ and the unwelcome $x$. We subtract it away:

$$\frac{x^3}{x^2+1} = \frac{(x^2+1)\cdot x - x}{x^2+1} = \frac{(x^2+1)\cdot x}{x^2+1} - \frac{x}{x^2+1} = x - \frac{x}{x^2+1}.$$

Therefore

$$\int \frac{x^3}{x^2+1} dx = \int \left(x - \frac{x}{x^2+1}\right) dx = \frac{x^2}{2} - \frac{1}{2}\ln\left(x^2+1\right) + C.$$

**Example 92** Let $g(x) = \frac{x^4}{x^4-1}$. It is sufficient to notice that $x^4 = \left(x^4 - 1\right) + 1$. Therefore

$$\frac{x^4}{x^4-1} = \frac{\left(x^4-1\right)+1}{x^4-1} = \frac{x^4-1}{x^4-1} + \frac{1}{x^4-1} = 1 + \frac{1}{x^4-1}.$$

### 6.3.1 How to write a proper rational function as a sum of partial fractions

Integrating a polynomial is straightforward, so let us suppose that we are given a proper rational function. We would like to represent it as a sum of so called **partial fractions**. Without being overly formal, let us outline the procedure of partial fraction decomposition.

There are two basic types of partial fractions:

$$\text{Type I:} \qquad \frac{A}{(ax+b)^n},$$

$$\text{Type II:} \qquad \frac{Ax+B}{(ax^2+bx+c)^n}.$$

In principle, any rational function can be written as a sum of a polynomial and fractions of the form above. The first step is to factor the denominator $q(x)$. Theoretically, it is always possible to write a polynomial as a product of linear terms and irreducible quadratics (it follows from the so-called fundamental theorem of algebra). In practice, this can be quite a challenge, as discussed in Section 1.5.

Second, we write $\frac{p(x)}{q(x)}$ as a sum of partial fractions with undetermined coefficients. A factor in the form $(ax+b)^n$ in the denominator $q(x)$ leads to $n$ fractions:

$$\frac{A_1}{ax+b}, \frac{A_2}{(ax+b)^2}, \ldots, \frac{A_n}{(ax+b)^n}.$$

A factor in the form $\left(ax^2+bx+c\right)^n$ also leads to $n$ fractions:

$$\frac{B_1x+C_1}{ax^2+bx+c}, \frac{B_2x+C_2}{(ax^2+bx+c)^2}, \ldots, \frac{B_nx+C_n}{(ax^2+bx+c)^n}.$$

The third step is to find the coefficients.

**Example 93** In Example 89 we used the method of Divine Inspiration to rewrite the expression $\frac{2}{x^2-1}$. Let us now do it systematically. Start by factoring $x^2 - 1 = (x+1)(x-1)$. The partial fractions needed are $\frac{A}{x+1}$ and $\frac{B}{x-1}$. We would like to have

$$\begin{aligned} \frac{2}{x^2-1} &= \frac{A}{x+1} + \frac{B}{x-1} \\ &= \frac{A(x-1)}{(x+1)(x-1)} + \frac{B(x+1)}{(x+1)(x-1)} = \frac{A(x-1)+B(x+1)}{(x+1)(x-1)}, \end{aligned}$$

which will happen as long as $A(x-1) + B(x+1) = 2$. The two polynomials are equal if and only if their corresponding coefficients match. We have

$$A(x-1) + B(x+1) = Ax - A + Bx + B = (A+B)x + (B-A),$$

which means that the coefficient in front of $x$ is $A+B$ and the free term is $B-A$. Since the polynomial in the other numerator is just $2 = 0 \cdot x + 2$, we need

$$\begin{cases} A + B = 0, \\ B - A = 2. \end{cases}$$

Solving for $A$ and $B$, we get $A = -1$ and $B = 1$.

**Example 94** Another method to find the coefficients $A$ and $B$ from the previous example is to observe that if two polynomials are equal, then they must give the same values for all values of $x$. Since the equality $2 = A(x-1) + B(x+1)$ is supposed to hold for all $x$, in particular it must be true for some carefully chosen values of $x$. We have:

$$\begin{aligned} 2 &= 2B, \quad \text{for } x = 1, \\ 2 &= -2A, \quad \text{for } x = -1. \end{aligned}$$

Hence, $A = -1$ and $B = 1$.

The two methods illustrated in Examples 93 and 94 can be combined for an efficient calculation of the unknown coefficients.

**Example 95** Since $x^4 - 1 = (x+1)(x-1)(x^2+1)$, we get the partial fraction decomposition of

$$\frac{1}{x^4-1} = \frac{A}{x+1} + \frac{B}{x-1} + \frac{Cx+D}{(x^2+1)}.$$

Adding the fractions and equating the numerators on both sides we get

$$1 = A(x-1)\left(x^2+1\right) + B(x+1)\left(x^2+1\right) + (Cx+D)\left(x^2-1\right).$$

There are two obvious values of $x$ which are good to plug in: $1$ and $-1$. This gives

$$\begin{aligned} 1 &= 4B, \quad \text{for } x = 1, \\ 1 &= -4A, \text{ for } x = -1, \end{aligned}$$

hence $A = -\frac{1}{4}$ and $B = \frac{1}{4}$. Plugging in $x = 0$ is fairly easy and it gives

$$1 = A(-1) + B + D(-1) = -\frac{1}{4}(-1) + \frac{1}{4} - D = \frac{1}{2} - D,$$

so $D = -\frac{1}{2}$. Finally, let us notice that if we multiplied the whole thing out, the coefficient in front of $x^3$ would be $A + B + C = 0$, as there is no $x^3$ on the left-hand side, which is a constant polynomial 1. This gives $C = 0$. Putting it all together we get

$$\frac{1}{x^4-1} = \frac{1}{4(x-1)} - \frac{1}{4(x+1)} - \frac{1}{2(x^2+1)}.$$

**Example 96** (for readers who have heard of complex numbers) Another option in the previous example would be to put $x = i = \sqrt{-1}$. This would give

$$1 = A \cdot 0 + B \cdot 0 + (C \cdot i + D)(-2) = -2C \cdot i - 2D,$$

and since two complex numbers are equal whenever their real and imaginary parts are equal, we immediately get $-2C = 0$ and $-2D = 1$, which gives $C = 0$ and $D = -\frac{1}{2}$. That's two coefficients for the price of one!

**Example 97** Consider $f(x) = \frac{3x^3-6x^2-14x+20}{x^3(x-2)}$. We need to rewrite the function as a sum of the form $\frac{A}{x} + \frac{B}{x^2} + \frac{C}{x^3} + \frac{D}{x-2}$. (Why do we need $\frac{A}{x}$ and $\frac{B}{x^2}$? Again, we would like the expression as simple as possible, yet complicated enough to give us the desired function. The expression $\frac{C}{x^3} + \frac{D}{x-2} = \frac{Dx^3+Cx-2C}{x^3(x-2)}$ is not sufficiently flexible. Part of the trouble is, no matter how hard we try, we are not going to get the term $-6x^2$ needed in the numerator.) We add the fractions:

$$f(x) = \frac{Ax^2(x-2) + Bx(x-2) + C(x-2) + Dx^3}{x^3(x-2)}.$$

Plugging $x = 0$ and $x = 2$ into both numerators, we get

$$\begin{aligned} 20 &= -2C, \\ -8 &= 8D, \end{aligned}$$

which gives $C = -10$ and $D = -1$. Unfortunately, there are no other nice values of $x$ to plug in. Without really multiplying out the numerator, we can compare the coefficients in front of $x^3$:

$$3 = A + D.$$

It follows that $A = 3 - D = 4$. Having no better options, we plug in a randomly chosen value of $x$, say $x = 1$:

$$3 = -A - B - C + D,$$

hence $B = -3 - A - C + D = -3 - 4 + 10 - 1 = 2$ and

$$\frac{3x^3 - 6x^2 - 14x + 20}{x^3(x-2)} = \frac{4}{x} + \frac{2}{x^2} - \frac{10}{x^3} - \frac{1}{x-2}.$$

### 6.3.2 How to integrate partial fractions

We have seen several examples already, but let us summarize what we have seen. For partial fractions of Type I we have

$$\int \frac{A}{(ax+b)^n} dx = \begin{cases} \frac{A}{a} \ln|ax+b|\, dx & \text{if } n = 1, \\ \frac{A}{a(1-n)(ax+b)^{n-1}} & \text{if } n > 1. \end{cases}$$

There is no need to memorize this formula. One can always make a substitution $u = ax + b$ and integrate $\int u^{-n} du = \frac{1}{-n+1} u^{-n+1} + C = \frac{1}{(1-n)u^{n-1}} + C$ (for $n \neq 1$).

Type II antiderivatives can get quite messy in general. First, one needs to complete the squares in the denominator and perform the appropriate substitution. For $n = 1$, after completing the squares and "shifting the variable," we find the answer as a sum of two functions, one involving natural logarithm and the other inverse tangent. For $n \geq 2$, we use the following reduction formula:

$$\int \frac{dx}{(ax^2+c)^n} = \frac{x}{2(n-1)c(ax^2+c)^{n-1}} + \frac{2n-3}{2(n-1)c} \int \frac{dx}{(ax^2+c)^{n-1}}.$$

At each step the power of the denominator is decreased by 1. (Luckily, in most problems we do not need to resort to such unpleasant measures.)

**Example 98** Following Examples 92 and 95 we have

$$\begin{aligned} \int \frac{x^4}{x^4-1} dx &= \int \left(1 + \frac{1}{x^4-1}\right) dx \\ &= \int \left[1 + \frac{1}{4(x-1)} - \frac{1}{4(x+1)} - \frac{1}{2(x^2+1)}\right] dx \\ &= x + \frac{1}{4} \ln|x-1| - \frac{1}{4} \ln|x+1| - \frac{1}{2} \arctan x + C. \end{aligned}$$

**Example 99** Let us integrate the function from Example 97

$$\begin{aligned} \int \frac{3x^3 - 6x^2 - 14x + 20}{x^3(x-2)} dx &= \int \left(\frac{4}{x} + \frac{2}{x^2} - \frac{10}{x^3} - \frac{1}{x-2}\right) dx = \\ &= 4 \ln|x| - \frac{2}{x} + \frac{5}{x^2} - \ln|x-2| + C. \end{aligned}$$

## 6.4 Trigonometric antiderivatives

In the previous section we learned to integrate any rational function. We would also like to integrate functions involving square roots. Unfortunately, there is no general method to integrate $\sqrt{f(x)}$, even if we know how to integrate $f(x)$.

The good news is that, at least sometimes, we are able to integrate expressions involving $\sqrt{a^2 - x^2}$, $\sqrt{x^2 - a^2}$ and $\sqrt{x^2 + a^2}$ (where $a$ is a constant). One of the main reasons we are interested in integrals involving trigonometric functions is that they are useful in integrating expressions of this sort.

**Example 100** In order to integrate $\int \frac{1}{\sqrt{1-x^2}} dx$, it is helpful to make a substitution $x = \sin t$. (OK, so far we made substitutions of the form $u = u(x)$, rather than $x = x(t)$. One could write $t = \arcsin(x)$ instead of $x = \sin t$, but it is convenient to do the latter.) We have

$$1 - x^2 = 1 - \sin^2 t = \cos^2 t,$$

which allows us to replace $\sqrt{1 - x^2}$ with $\cos t$ (at least assuming that $t$ is in the first quadrant). It is essential not to forget that $dx = \cos t dt$. We have

$$\int \frac{1}{\sqrt{1 - x^2}} dx = \int \frac{1}{\cos t} \cdot \cos t dt = \int 1 dt = t + C,$$

and since $t = \arcsin(x)$, the answer to our integration problem is

$$\int \frac{1}{\sqrt{1 - x^2}} dx = \arcsin(x) + C.$$

Of course we could have recognized that, (see Section 3.4), but the method works for more complicated integrals as well.

The substitution $x = \sin t$ may not be the most natural thing to do, but it works well, as it allows for the expression under the square root to be written as a square of something fairly simple. In general, the trigonometric substitutions are as follows.

| $f(x)$ under $\sqrt{\phantom{x}}$ | Substitution | $\sqrt{f(x)}$ | $dx$ |
|---|---|---|---|
| $a^2 - x^2$ | $x = a \sin t$ | $a \cos t$ | $a \cos t dt$ |
| $x^2 - a^2$ | $x = a \sec t$ | $a \tan t$ | $a \sec t \tan t dt$ |
| $a^2 + x^2$ | $x = a \tan t$ | $a \sec$ | $a \sec^2 t dt$ |

(6.10)

The reader can easily verify the formulas for $\sqrt{f(x)}$ and for $dx$ in the table above. A frequent question students ask is "how did you know to make that substitution?" For most of us the answer should be "because someone showed me that it works." It is hard to say how long it would take any of us to come up with this idea. However, it should be clear why it works: whatever was under the radical can now be written as a square.

### 6.4.1 Instead of the square root we have trigonometric functions. Now what?

Converting an integral to a trigonometric one is only helpful if we can actually integrate the new functions. For this reason, it is useful to brush up on some methods and tricks to integrate various trigonometric integrals. Let us start with a few examples:

**Example 101** $\int \sin^3 x \cos x dx = \frac{1}{4}\sin^4 x + C$. We can obtain this answer by substituting $u = \sin x$, which gives $du = \cos x dx$ and $\int \sin^3 x \cos x dx$ $= \int u^3 du = \frac{1}{4}u^4 + C$.

**Example 102** $I = \int \cos^3 x dx = \int \cos^2 x \cos x dx = \int \left(1 - \sin^2 x\right) \cos x dx$. Put $u = \sin x$ to get $du = \cos x dx$ and $I = \int \left(1 - u^2\right) du = -\frac{1}{3}u^3 + u + C$ $= -\frac{1}{3}\sin^3 x + \sin x + C$.

**Example 103** $I = \int \sin^3 x dx = \int \sin^2 x \sin x dx = \int \left(1 - \cos^2 x\right) \sin x dx$. Put $u = \cos x$ to get $du = -\sin x dx$ and $I = \int \left(1 - u^2\right)(-1) du = \int \left(u^2 - 1\right) du$ $= \frac{1}{3}u^3 - u + C = \frac{1}{3}\cos^3 x - \cos x + C$.

**Example 104** How can we integrate $\tan x$? Substituting $u = \cos x$ we get $du = -\sin x dx$, hence

$$\begin{aligned} \int \tan x dx &= \int \frac{\sin x}{\cos x} dx = -\int \frac{1}{u} du \\ &= -\ln|u| + C = -\ln|\cos x| + C. \end{aligned} \tag{6.11}$$

**Example 105** To integrate $I = \int \frac{\cos^3 x}{\sin^2 x} dx$ we substitute $u = \sin x$. Then $du = \cos x dx$ and we get $I = \int \frac{\cos^2 x}{\sin^2 x} \cos x dx = \int \frac{1-u^2}{u^2} du = \int \left(\frac{1}{u^2} - 1\right) du$ $= -\frac{1}{u} - u + C = -\frac{1}{\sin x} - \sin x + C$.

These simple substitutions help us integrate $\int \sin^m x \cos^n x dx$, where $m$ and $n$ are integers, as long as at least one of them is odd. Let us say $m$ is odd. We take $u = \cos x$. The reason this works is that $du = \sin x$ "uses" one of the sines. What is left is an even power of $\sin x$, which can always be written in terms of $\cos x$ without taking square roots, just as in Example 101. If $n$ is odd, then we take $u = \sin x$ as in Example 102.

To integrate products or quotients of sine and cosine with both powers even is a little harder. We need to use formulas (1.9) from Section 1.7. Specifically, since $\cos(2x) = 2\cos^2 x - 1 = 1 - 2\sin^2 x$, we have

$$\begin{aligned} \cos^2 x &= \frac{1}{2}(1 + \cos(2x)), \qquad (6.12) \\ \sin^2 x &= \frac{1}{2}(1 - \cos(2x)). \end{aligned}$$

**Example 106**

$$\begin{aligned} I &= \int \sin^4 x \cos^2 x dx = \int \left(\sin^2 x\right)^2 \cos^2 x dx \\ &= \int \left(\frac{1}{2}(1 - \cos(2x))\right)^2 \frac{1}{2}(1 + \cos(2x)) dx \\ &= \frac{1}{8}\int \left[\cos^3(2x) - \cos^2(2x) - \cos(2x) + 1\right] dx. \end{aligned}$$

The first term $\cos^3(2x)$ can now be handled by substitution $u = \sin(2x)$, as in Example 102, leading to $\int \cos^3(2x) dx = -\frac{1}{6}\sin^3(2x) + \frac{1}{2}\sin(2x) + C$. The second term needs to be rewritten using the half-angle formulas (6.12) again: $\cos^2(2x) = \frac{1}{2}(1 + \cos(4x))$, which gives $\int \cos^2(2x) dx = \frac{1}{2}x + \frac{1}{8}\sin(4x) + C$. The rest is left to the reader as an exercise.

Another type of trigonometric integral which occurs frequently is in the form $\int \tan^m x \sec^n x dx$. A technique similar to the one described for expressions involving sine and cosine can be used. If $n$ is even, we substitute $u = \tan x$. If $m$ is odd, we substitute $u = \sec x$.

**Example 107** To integrate $I = \int \tan^3 x \sec^2 x dx$, substitute $u = \tan x$, which gives $du = \sec^2 x dx$. Hence $I = \int u^3 du = \frac{1}{4}u^4 + C = \frac{1}{4}\tan^4 x + C$.

**Example 108** We can look at the integral from Example 104 one more time: $I = \int \tan x dx$. We use $u = \sec x$ with $du = \sec x \tan x dx$. Next:

$$\begin{aligned} I &= \int \frac{1}{\sec x} \sec x \tan x dx = \int \frac{1}{u} du \qquad (6.13) \\ &= \ln |u| + C = \ln |\sec x| + C, \end{aligned}$$

which is not a contradiction with formula (6.11) as $\ln |\sec x| = \ln \frac{1}{|\cos x|} = \ln 1 - \ln |\cos x| = -\ln |\cos x|$.

**Example 109** This time let $I = \int \tan^3 x \sec^3 x dx$. Let $u = \sec x$. Then $du = \sec x \tan x dx$. We can "borrow" $\tan x \sec x$ from the integrand and write $I = \int \left(\tan^2 x \sec^2 x\right) (\sec x \tan x) dx$. Since $\tan^2 x = \sec^2 x - 1$, we have

$$\begin{aligned} I &= \int \left(u^2 - 1\right) u^2 du = \int \left(u^4 - u^2\right) du \\ &= \frac{1}{5} u^5 - \frac{1}{3} u^3 + C = \frac{1}{5} \sec^5 x - \frac{1}{3} \sec^3 x + C. \end{aligned}$$

We have more work to do in the unlucky situation when $m$ is even and $n$ is odd. We begin with just plain $\sec x$, integrating which is a little tricky, but not very hard. We multiply and divide the integrand by the sum $\sec x + \tan x$:

$$\begin{aligned} \int \sec x dx &= \int \sec x \frac{\sec x + \tan x}{\sec x + \tan x} dx \\ &= \int \frac{\sec^2 x + \sec x \tan x}{\sec x + \tan x} dx. \end{aligned}$$

It becomes clear why this awkward trick helps once we notice that the numerator is now the derivative of the denominator. Hence (see formula (6.3)),

$$\int \sec x dx = \ln |\sec x + \tan x| + C \qquad (6.14)$$

In the general case of an even power of $\tan x$ and an odd power of $\sec x$, the integrand can be represented entirely in terms of $\sec x$, since $\tan^2 x = \sec^2 x - 1$. Next, we can reduce the power $\sec^n x$ using a reduction formula (6.8).

**Example 110** Using (6.8) twice we have

$$\begin{aligned}
\int \sec^5 x dx &= \frac{\sec^3 x \tan x}{4} + \frac{3}{4}\int \sec^3 x dx \\
&= \frac{\sec^3 x \tan x}{4} + \frac{3}{4}\left[\frac{\sec x \tan x}{2} + \frac{1}{2}\int \sec x dx\right] \\
&= \frac{\sec^3 x \tan x}{4} + \frac{3\sec x \tan x}{8} + \frac{3}{8}\ln\left|\sec x + \tan x\right| + C.
\end{aligned}$$

Finally, let us use the trigonometric substitutions (6.10) to integrate some functions we could not handle before.

**Example 111** Let $I = \int \sqrt{4-x^2}dx$. We substitute $x = 2\sin t$, which gives $dx = 2\cos t dt$. Also, $4 - x^2 = 4 - 4\sin^2 t = 4\left(1-\sin^2 t\right) = 4\cos^2 t$. We get

$$\begin{aligned}
I &= \int \sqrt{4\cos^2 t}\,(2\cos t)\,dt = \int (2\cos t)\,(2\cos t)\,dt \\
&= \int 4\cos^2 t dt = \int 2\,(1+\cos 2t)\,dt = 2t + \sin 2t + C.
\end{aligned}$$

We now need to express the answer in terms of the variable $x$. Since $x = 2\sin t$, then $t = \arcsin(x/2)$. It also helps to notice that $\sin 2t = 2\sin t\cos t$. We know that $2\sin t = x$ and $4\cos^2 t = 4 - x^2$, hence $\cos t = \frac{1}{2}\sqrt{4-x^2}$. Putting this all together we obtain:

$$I = 2\arcsin(x/2) + x\frac{\sqrt{4-x^2}}{2} + C.$$

Some of us like to draw a triangle to help express various trigonometric functions in terms of $x$. In this example, we would start by drawing a right triangle with hypotenuse of length 2 and opposite side of length $x$. It follows from Pythagorean theorem that the adjacent side is length $\sqrt{4-x^2}$.

**Example 112** A different substitution is needed for the integral $I = \int \frac{1}{\sqrt{x^2-4}}dx$. This time we need $x = 2\sec t$, which gives $dx = 2\sec t \tan t dt$ and $x^2 - 4 = 4\sec^2 t - 4 = 4\tan^2 t$. Hence

$$\begin{aligned} I &= \int \frac{1}{\sqrt{4\tan^2 t}} 2\sec t \tan t dt \\ &= \int \frac{1}{2\tan t} 2\sec t \tan t dt = \int \sec t dt = \frac{1}{2}\ln|\sec t + \tan t| + C \\ &= \frac{1}{2}\ln\left|\frac{x}{2} + \frac{\sqrt{x-4^2}}{2}\right| + C = \frac{1}{2}\ln\left|x + \sqrt{x-4^2}\right| + \overline{C}, \end{aligned}$$

(as $\ln\left|\frac{x}{2} + \frac{\sqrt{x-4^2}}{2}\right| = \ln\left|x + \sqrt{x-4^2}\right| - \ln 2$, and $\ln 2$ is just another constant).

**Example 113** To illustrate the use of the third substitution (6.10) let us integrate $I = \int \sqrt{1+x^2}dx$. Take $x = \tan t$ with $dx = \sec^2 t dt$. We get $1+x^2 = 1+\tan^2 t = \sec^2 t$ and

$$\begin{aligned} I &= \int \sqrt{\sec^2 t}\sec^2 t dt = \int \sec^3 t dt \\ &= \frac{\sec t \tan t}{2} + \frac{1}{2}\int \sec t dt \\ &= \frac{\sec x \tan x}{2} + \frac{1}{2}\ln|\sec t + \tan t| + C = \\ &= \frac{1}{2}x\sqrt{x^2+1} + \frac{1}{2}\ln\left(x + \sqrt{x^2+1}\right) + C. \end{aligned}$$

## 6.5 Integration by an "educated guess"

Having calculated a few integrals, we can develop some intuition regarding the type of an answer we should expect. Some integrals can be "guessed," perhaps up to some constants. This leads to the method of an "educated guess," properly known as the **method of undetermined coefficients**. Although not very popular among students, this method is quite simple and elegant.

**Example 114** The integral from Example 80 can be predicted to look like $\int xe^x dx = (ax+b)\,e^x$. That is, we need to find coefficients $a$ and $b$ such that

$$xe^x = \frac{d}{dx}\left((ax+b)\,e^x\right) = ae^x + (ax+b)\,e^x = e^x\,(ax+a+b)\,,$$

which will happen if $a=1$ and $a+b=0$. This gives $b=-1$ and

$$\int xe^x dx = (x-1)\,e^x + C.$$

A similar method can be used to predict integrals of the form

$$\int (A\sin x + B\cos x)\,e^{kx} dx = (a\sin x + b\cos x)\,e^{kx} + C.$$

**Example 115** We predict

$$\int e^{2x}\cos x dx = (a\sin x + b\cos x)\,e^{2x} + C.$$

Differentiating the right-hand side by product rule and simplifying we get

$$\begin{aligned}
e^{2x}\cos x &= \frac{d}{dx}\left((a\sin x + b\cos x)\,e^{2x}\right) \\
&= e^{2x}\,(a\cos x - b\sin x + 2b\cos x + 2a\sin x) \\
&= e^{2x}\,((2a-b)\sin x + (a+2b)\cos x)\,;
\end{aligned}$$

matching the corresponding coefficients leads to

$$\begin{aligned}
2a - b &= 0, \\
a + b &= 1;
\end{aligned}$$

so $a=\frac{1}{3}$ and $b=\frac{2}{3}$; hence $\int e^{2x}\cos x dx = \frac{1}{3}\,(\sin x + 2\cos x)\,e^{kx} + C$. (Why do we need the $\sin x$ in the predicted answer? Without it, we would never get rid of the $2a\sin x$.)

## 6.6 Problems

1. Find the integrals:

   (a) $\int_0^1 x\sqrt{x^2+5}dx$

   (b) $\int \arctan x dx$

   (c) $\int x^2 e^x dx$ (by parts or by the method of undetermined coefficients)

   (d) $\int \frac{6x+5}{x^2+1}dx$

   (e) $\int \frac{1}{1-\sqrt{x}}dx$ (start by substituting $u = \sqrt{x}$)

2. Find the integral $\int \sec x dx$ by noticing that $\sec x = \frac{1}{\cos x} = \frac{\cos x}{\cos^2 x} = \frac{\cos x}{1-\sin^2 x}$ and then using the substitution $u = \sin x$.

3. Integrate $\int \frac{1}{x^2\sqrt{x^2-4}}dx$. Note that the answer does not contain any trigonometric functions after all.

4. Find $\int \frac{1}{(x^2+1)^2}dx$ using the trigonometric substitution $x = \tan t$.

5*. Let $I_n = \int \frac{1}{(x^2+1)^n}dx$. Use integration by parts with $u = \frac{1}{(x^2+1)^n}$ and $dv = dx$ to obtain the recursion formula: $I_{n+1} = \frac{1}{2n}\frac{x}{x^2+1} + \frac{2n-1}{2n}I_n$. Next, express $I_n$ in terms of $I_{n-1}$. Use the formula to find, for example, $\int \frac{1}{(x^2+1)^2}dx$.

6*. One way to define the natural logarithm function is to say that by definition $\ln(x) = \int_1^x \frac{1}{t}dt$. To get some flavor of this argument, let us say that $L(x) = \int_1^x \frac{1}{t}dt$. Use the properties of integrals to show that $L(ab) = L(a) + L(b)$. (Do not use logarithms, as we are pretending to have never heard of them.)

# Chapter 7

# Selected Applications of the Integral

## 7.1 Arc length

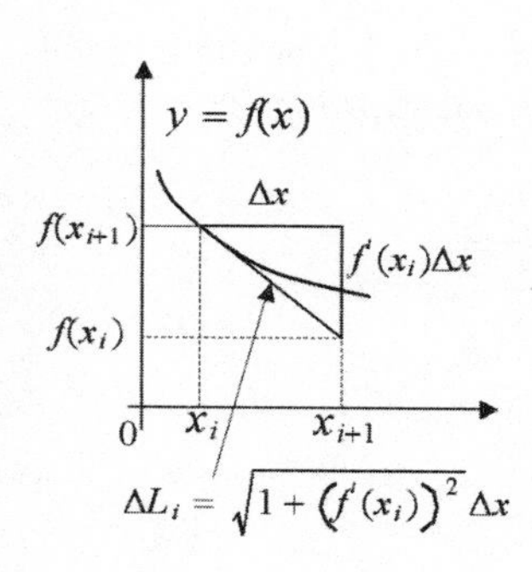


Suppose we would like to calculate the arc length of a curve which is a graph of some function $y = f(x)$ over an interval $[a, b]$. Let us divide the interval into several short subintervals by $n + 1$ evenly spaced points $a = x_0 < x_1 < \cdots < x_{n-1} < x_n = b$. In other words, $x_i = a + \Delta x \cdot i$, where $\Delta x = x_i - x_{i-1}$. For a given value of $x_i$ we can approximate the arc length of the short segment of the graph connecting the points $P = (x_i, f(x_i))$ and $Q = (x_i + \Delta x, f(x_i + \Delta x))$ by replacing the actual function with its linear approximation, centered at the point $(x_i, f(x_i))$. The tangent line approximating $f$ near $x = x_i$ has slope $f'(x_i)$; hence $f(x) \approxeq f(x_i) + f'(x_i)(x - x_i)$ (see Section 4.4). The length of the straight line segment connecting $P$ and $Q$ can be calculated using the Pythagorean theorem:

$$\Delta L_i = \sqrt{(\Delta x)^2 + (f'(x_i) \cdot \Delta x)^2} = \sqrt{1 + (f'(x_i))^2}\Delta x. \tag{7.1}$$

Using formula (7.1) on each of the line segments of the graph, we approximate the arc length

$$L \approxeq \sum_{i=0}^{n-1} \sqrt{1 + (f'(x_i))^2}\Delta x. \tag{7.2}$$

The expression in formula (7.2) is exactly the Riemann sum with $n$ equal subdivisions for a function $g(x) = \sqrt{1 + (f'(x))^2}$ on $[a, b]$ (see Section 5.4). Taking the limit of the sum as $n \to \infty$, we obtain

$$L = \int_a^b \sqrt{1 + (f'(x))^2}dx. \tag{7.3}$$

The good news is that we obtain a nice, simple formula for the arc length. The bad news is that the antiderivative of $\sqrt{1 + (f'(x))^2}$ typically turns out to be impossible to find, except for problems specifically designed to illustrate the arc length formula. In reality we must often resort to numerical computation of the integral, which brings us back to Riemann sums in some form or another.

**Example 116** Let us find the arc length of the graph of $y = x^2$ over the interval $[0, 1]$. Even in this simple case the integration requires substituting $2x = \tan t$ (see Section 6.4 for more details). We have

$$\begin{aligned} L &= \int_0^1 \sqrt{1 + (f'(x))^2}dx = \int_0^1 \sqrt{1 + (2x)^2}dx \\ &= \left[\frac{1}{4}\ln\left(2x + \sqrt{4x^2+1}\right) + \frac{1}{2}x\sqrt{4x^2+1}\right]\Bigg|_0^1 \\ &= \frac{1}{4}\ln\left(\sqrt{5}+2\right) + \frac{1}{2}\sqrt{5}. \end{aligned}$$

**Example 117** This example shows a typical trick needed in problems about arc length, which gives us another excuse to refresh the technique of completing the square. Consider a function $f(x) = x^2 - \frac{1}{8}\ln x$ on the interval $[1, e]$. We have $f'(x) = 2x - \frac{1}{8x}$. Furthermore, we get

$$\begin{aligned} 1 + \left(f'(x)\right)^2 &= 1 + \left(2x - \frac{1}{8x}\right)^2 = 1 + \frac{1}{64x^2} + 4x^2 - \frac{1}{2} \\ &= \frac{1}{64x^2} + 4x^2 + \frac{1}{2} = \left(2x + \frac{1}{8x}\right)^2, \end{aligned}$$

which works out because the coefficients in the function were carefully selected to obtain $-\frac{1}{2}$, which was subsequently replaced by $\frac{1}{2}$, after adding 1. As a result, the expression $1 + \left(f'(x)\right)^2 = \frac{1}{64x^2} + 4x^2 + \frac{1}{2}$ becomes a complete square, thus making it feasible to calculate

$$\begin{aligned} L &= \int_1^e \sqrt{1 + (f'(x))^2} dx = \int_1^e \sqrt{\left(2x + \frac{1}{8x}\right)^2} dx = \\ &= \int_1^e \left(2x + \frac{1}{8x}\right) dx = \left[x^2 + \frac{1}{8}\ln x\right]\Big|_1^e = e^2 - \frac{7}{8}. \end{aligned}$$

**Exercise 118** *The arc length formula (7.3) can also be obtained using an alternative approach. For a function $f(x)$ on an interval $[a, b]$, define $L(x)$ as the arc length of the curve $y = f(x)$ over the interval $[a, x]$. Show that the derivative $L'(x) = \sqrt{1 + (f'(x))^2}$. It then follows that $L(x)$ is an antiderivative of $\sqrt{1 + (f'(x))^2}$, that is, $L(x) = \int_a^b \sqrt{1 + (f'(x))^2} dx + C$, where $C$ is some unknown constant. However, $L(a) = 0$, by the definition of $L(x)$, so $C$ must equal 0. (The reader should not get frustrated as this exercise is relatively difficult. It is perfectly safe to skip it and to continue reading.)*

## 7.2 Area

The integral was defined in Section 5.1 as the (signed) area between the graph of a function and the $x$-axis over some interval. Integrals can also be used to find the area between curves.

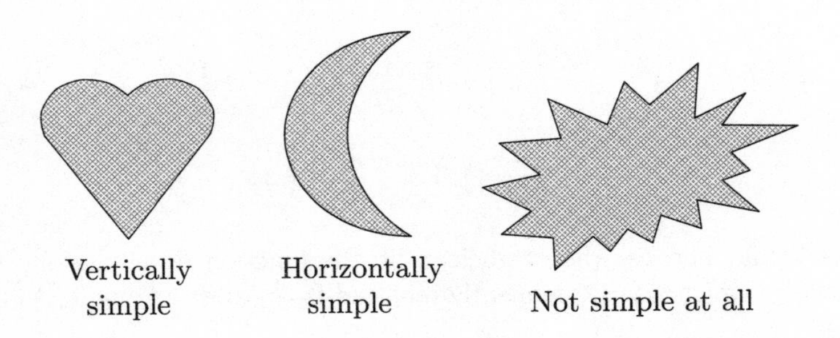


Some regions in the plane (typically called **vertically simple**) can be described as bounded above and below by graphs of two functions of $x$. It may also be possible to describe a region as lying between graphs of two functions of $y$ (a **horizontally simple** region). In both cases the area can be calculated as the integral

$$A = \int_a^b Larger\,(t) - Smaller\,(t)\, dt.$$

For a vertically simple region we use integration with respect to $x$, and for a horizontally simple region we use integration with respect to $y$. If the variable of integration is $x$, then "Larger" means a function whose graph is higher up. When the variable is $y$, then "Larger" refers to the graph farther to the right. One of the difficulties in solving problems about areas is finding the intersection points of the given curves. While the official formula for the area bounded by two curves is frequently given as $\int_a^b |f(t) - g(t)|\, dt$, we still need to find the points of intersection and divide the interval of integration into appropriate subintervals on which one of the functions is larger than the other.

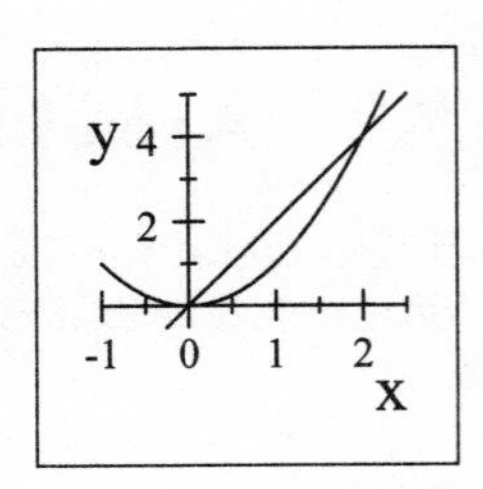


**Example 119** Find the area bounded by the graphs of the functions $f(x) = x^2$ and $g(x) = 2x$. The first step is to find the intersection points of the two graphs. We can do that by solving the equation $x^2 = 2x$. Moving $x$ to the left-hand side and factoring, we get $x(x-2) = 0$, which tells us that

the intersection points correspond to $x = 0$ and $x = 2$. We have

$$A = \int_0^2 \left(2x - x^2\right) dx = \left(x^2 - \frac{1}{3}x^3\right)\Big|_0^2 = 4 - \frac{8}{3} = \frac{4}{3}.$$

**Example 120** Let us calculate the area of the same region, this time treating it as horizontally simple. The curve $y = x^2$, which was the lower one, is now the one farther to the right. We describe it as $x = f^{-1}(y) = \sqrt{y}$. The "curve" $y = 2x$ is now farther to the left, and we also view it as a function of $y$, that is, $x = g^{-1}(y) = \frac{1}{2}y$. The intersection points correspond to $y = 0$ and $y = 4$. It takes just a little more effort to get the same answer:

$$\begin{aligned} A &= \int_0^4 \left(\sqrt{y} - \frac{1}{2}y\right) dy = \left(\frac{2}{3}y^{3/2} - \frac{1}{4}y^2\right)\Big|_0^4 \\ &= \frac{2}{3} \cdot 4^{3/2} - \frac{1}{4} \cdot 4^2 = \frac{2}{3} \cdot 8 - 4 = \frac{4}{3}. \end{aligned}$$

Clearly, not every region on the plane is both vertically and horizontally simple. When it is, sometimes there is a clear advantage in integrating with respect to one variable versus another. This is illustrated in the next example.

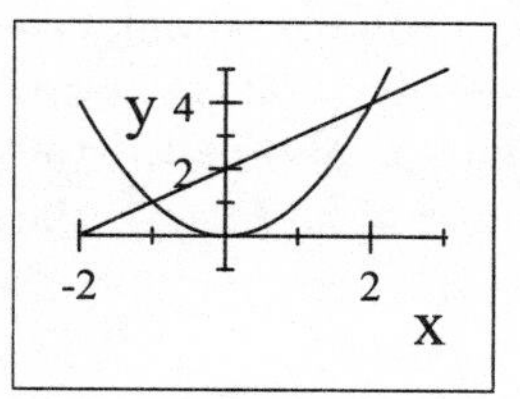


**Example 121** Find the area bounded by the graphs of the equations: $y = x^2$ and $y = x + 2$. The intersection points are easily found to be $(-1, 1)$ and $(2, 4)$. Integrating on the $x$-axis we obtain

$$A = \int_{-1}^2 \left((x + 2) - x^2\right) dx = \frac{9}{2}.$$

If we wish to calculate the same area by integrating on the $y$-axis from 0 to 4, we have to deal with a new difficulty: the formula for the "Smaller" function changes at $y = 1$. For $y \in [0, 1]$ the function is defined by $x = -\sqrt{y}$ and for $y \in [1, 4]$ we have $x = y - 2$. The "Larger" function is $x = \sqrt{y}$ for all $y \in [0, 4]$. We split the interval of integration and calculate:

$$\begin{aligned} A &= \int_0^1 [\sqrt{y} - (-\sqrt{y})]\, dy + \int_1^4 [\sqrt{y} - (y-2)]\, dy \\ &= \int_0^1 2\sqrt{y} dy + \int_1^4 (\sqrt{y} - y + 2)\, dy = \frac{4}{3} + \frac{19}{6} = \frac{9}{2}. \end{aligned}$$

## 7.3 Volumes of solids

Integrals can also be used to calculate the exact volume of certain solids. The method of **cross-sections** goes back to Archimedes, who used it to find the volume of a sphere some 2000 years before Newton (see [2], p. 334; an article by Max Den in [1], p. 33; and [5]). One way to look at the problem is to imagine slicing an oddly shaped block of cheese (I prefer to think about chocolate-covered marzipan) into several slices. If we know the area of a slice is $A(x)$, we can multiply it by its thickness $\Delta x$ and calculate its volume $\Delta V = A(x) \cdot \Delta x$. More formally, suppose that the area of a cross-section of a solid is given as a function of $x$ on an interval $[a, b]$. By arguments very similar to the ones from the previous section (involving Riemann sums), the volume of the solid can be calculated as

$$V = \int_a^b A(x)\, dx. \tag{7.4}$$

In particular, suppose that the solid is obtained by revolving about the $x$-axis a region bounded by the graph of $y = f(x)$ and the $x$-axis, above the interval $[a, b]$. In this case, the method of cross-sections is also called the method of **discs**. Since the area of a single disc $A(x) = \pi y^2 = \pi (f(x))^2$, the volume of the solid of revolution is computed as

$$V = \pi \int_a^b (f(x))^2\, dx. \tag{7.5}$$

Another frequent problem is to find the volume of a solid of revolution, where the rotated region is bounded by the graphs of two functions. In this case, a cross-section is a **washer**.

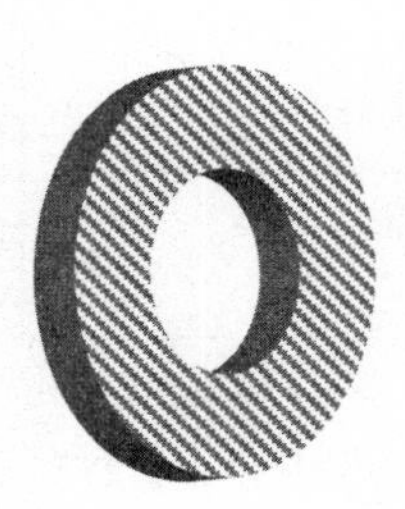

$\Delta V = A(x) \cdot \Delta x$

Its area can be calculated by subtracting the area of the smaller circle from the area of the larger one: $A(x) = \pi (f(x))^2 - \pi (g(x))^2$ (where $g$ is the smaller function). Therefore:

$$V = \pi \int_a^b (f(x))^2 - (g(x))^2 \, dx. \tag{7.6}$$

**WARNING:** One the most frequent mistakes in calculating the volume by integrating is using the square of the difference instead of the difference of the squares:
$V \neq \pi \int_a^b (f(x) - g(x))^2 \, dx$. This is just a bad idea.

**Example 122** The formula for the volume of a prism

$$V = \frac{1}{3} Ah, \tag{7.7}$$

where $A$ is the area of the base and $h$ is the height, can be derived without using calculus (see [6]), but we will use the method of cross-sections. In this example the cross-sections are not circular, as a prism is not a solid of revolution. Let us agree that the prism is "upside down" and that its base (say a rectangle) is parallel to the $xy$-plane. Consider slicing the prism with horizontal planes $z = const$. Using similarity of triangles we conclude that for a given value of $z \in [0, h]$ the area of the cross-section $A(z) = \left(\frac{z}{h}\right)^2 A$. (Checking this as a nice exercise for the reader.) Hence the volume

$$\begin{aligned} V &= \int_0^h A(z) \, dz = \int_0^h \left(\frac{z}{h}\right)^2 A dz \\ &= \frac{A}{h^2} \int_0^h z^2 dz = \frac{A}{h^2} \cdot \frac{1}{3} z^3 \bigg|_0^h = \frac{1}{3} Ah. \end{aligned}$$

The following two examples demonstrate the method of washers, as well as a different technique, the method of **cylindrical shells**, which is arguably best explained by an example.

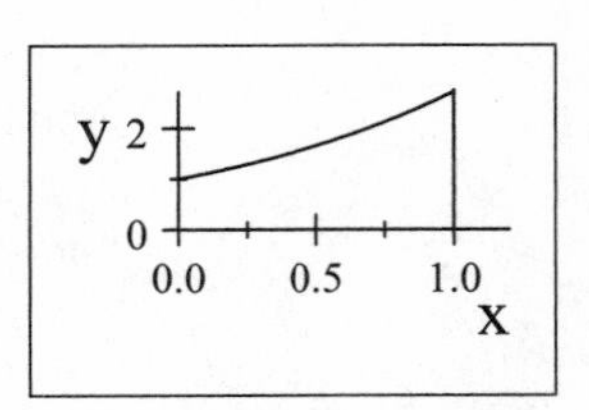


**Example 123** Let us find the volume of the solid obtained by rotating about the $y$-axis the region below the line $y = e^x$ and above the interval $[0, 1]$. We split the interval of integration on the $y$-axis $[0, e]$, since the cross-sections are discs for $y \in [0, 1]$ and washers for $y \in (1, e]$. The radius of the outside circle is always 1, but on the second interval we need to subtract the area of the smaller circle, whose radius is $x = \ln y$. We calculate:

$$\begin{aligned} V &= \pi \int_0^1 1 dy + \pi \int_1^e \left(1 - (\ln y)^2\right) dy \\ &= \pi + \pi \left[-y (\ln y - 1)^2\right]\Big|_1^e = \pi + \pi = 2\pi \end{aligned}$$

(the second integral takes some effort to evaluate, by parts, for example).

In the method of cylindrical shells the idea of slicing the solid with planes is replaced by "peeling" narrow cylindrical shells.

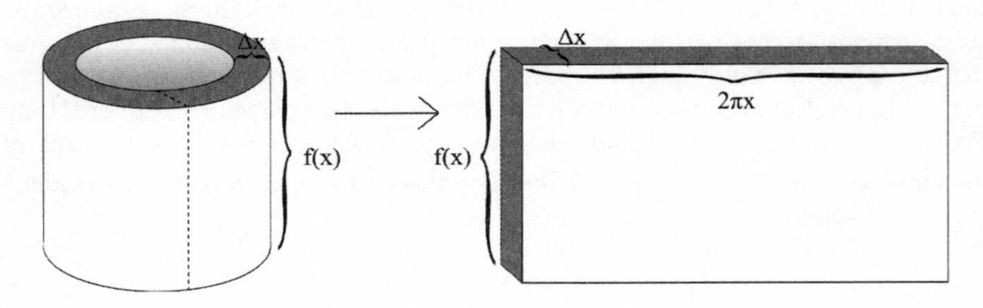


**Example 124** For every value of $x \in [0, 1]$, consider a cylindrical shell obtained by rotating, about the $y$-axis, a narrow rectangle above the interval $[x, x + \Delta x]$, with height $h = f(x) = e^x$. The volume of the shell can be well approximated by cutting it and laying it flat. This gives a rectangular box with base $2\pi x$ and width $f(x)$. The height of the box is $\Delta x$. The volume of the cylindrical shell is approximately $\Delta V \approxeq 2\pi x f(x) \Delta x$. By considering Riemann sums, as usual, and

taking the limit as the number of subdivisions goes to $\infty$, we obtain

$$V = 2\pi \int_0^1 xf(x)\,dx = 2\pi \int_0^1 xe^x dx = 2\pi \left[e^x (x-1)\right]|_0^1 = 2\pi.$$

The integration of $xe^x$ can be done by parts (see Example 80) or by the method of undetermined coefficients (Example 114).

## 7.4 Solving selected differential equations

Differential equations are used across the sciences to describe a wide variety of phenomena. They are equations involving the derivative of an unknown function $y(x)$. Frequently, we use $t$ instead of $x$ as the independent variable, as it often represents time. A solution of an algebraic equation is a number or a set of numbers. A solution of a differential equation is a function or a family of functions. In addition, we may require the solution to satisfy an initial condition, such as $y(0) = 5$ or $y(\pi) = 7$, for instance. We refer to a differential equation along with an initial condition(s) as an **initial value problem (IVP)**.

**Example 125** Consider an equation $\frac{dy}{dx} = 3y$. It is not hard to see that $y(x) = Ae^{3x}$ solves this equation, for every real value of the constant $A$. If we require in addition that $y(1) = 50$, then we have an IVP. Among all the functions in the form $y(x) = Ae^{3x}$, we choose the one for which $y(1) = Ae^3 = 50$. This gives $A = 50e^{-3} \approxeq 2.49$. This is of course an example of equation (3.4), which describes processes with the rate of growth (or decay) of $y$ proportional to the its current value.

A small variation of the previous example illustrates **Newton's law of cooling.**

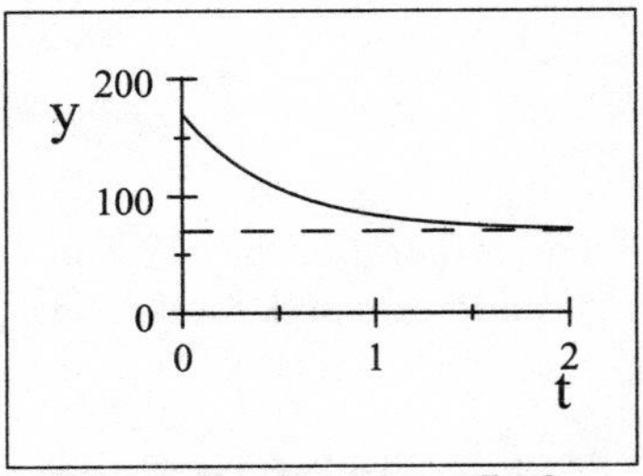

**Example 126** The rate of change of temperature $y'$ of a body is proportional to the difference between its temperature $y$ and the ambient temperature $T_0$:

$$y' = -k\,(y - T_0)\,. \tag{7.8}$$

How warm is your coffee?

It is easy to verify that all functions of the form

$$y\,(t) = T_0 + Ce^{-kt},$$

where $C$ is an arbitrary constant, solve equation (7.8). If we know in addition that the initial temperature is $T_1$, then the solution becomes $y\,(t) = T_0 + (T_1 - T_0)\,e^{-kt}$.

Assuming that the rate of growth $y'$ is always proportional to the value of $y$ does not always lead to the best model. The size of a population is usually limited by the size of a given environment, the availability of resources, etc. A more realistic approach is to introduce a so-called **carrying capacity**. This leads to the **logistic equation**:

$$\frac{dy}{dt} = ky\,(C - y)\,, \tag{7.9}$$

where the constant $C$ represents the largest size of the population which can be sustained. The same equation can also be used to describe an epidemic, where $C$ is the total size of the population and $y$ is the number of individuals infected. (A more cheerful application is to model a spreading rumor.)

**WARNING:** A common first impulse is to "integrate both sides." While there is nothing wrong with this idea in principle, one must pay attention to the variable of integration. The integral $\int \frac{dy}{dt}dt = y + const.$, which is why we wanted to integrate in the first place. However,

$$\int ky\,(C - y)\,dt \neq \frac{kC}{2}y^2 - \frac{k}{3}y^3 + const.,$$

since the variable of integration is $t$ and not $y$.

The trick is to divide both sides of (7.9) by $y\,(C-y)$:

$$\frac{1}{y\,(C-y)}\frac{dy}{dt}=k.$$

Now we can integrate both sides, using the substitution rule:

$$\int\left(\frac{1}{y\,(C-y)}\frac{dy}{dt}\right)dt=\int kdt,$$

which gives

$$\int\frac{dy}{y\,(C-y)}=kt+D.$$

We learned in Section 6.3 to rewrite the fraction

$$\frac{1}{y\,(C-y)}=\frac{1}{C}\left(\frac{1}{C-y}+\frac{1}{y}\right),$$

which gives

$$\int\frac{dy}{y\,(C-y)}=-\frac{1}{C}\left(\ln|C-y|-\ln|y|\right)+\overline{D}=-\frac{1}{C}\ln\left|\frac{C-y}{y}\right|+\overline{D}.$$

(Hang in there as this takes just a little more work). Since there is no point in adding an arbitrary constant to both sides of the equation, we obtain

$$-\frac{1}{C}\ln\left|\frac{C-y}{y}\right|=kt+D.$$

To isolate $y$ as a function of $t$, we multiply by $-C$ and apply the exponential function to both sides:

$$\left|\frac{C-y}{y}\right|=e^{-Ckt-DC}=Ae^{-Ckt},$$

where $A=e^{-DC}$ is an arbitrary positive number. We can drop the absolute value if we allow $A$ to be any real number. (Technically, we would never get $A=0$ out of $\pm e^{-DC}$, but allowing this value for $A$ does not lead to any trouble.) Multiply both sides by $y$ to get

$$C-y=Ae^{-Ckt}\cdot y,$$

and finally solve it for $y$,

$$y = \frac{C}{1 + Ae^{-Ckt}}.$$

Notice that this answer makes intuitive sense. For large values of $t$, the number $e^{-Ckt}$ becomes very close to 0 and $y(t) \approxeq C$. No matter what the initial value of $y$ was, eventually it should stabilize at about $C$.

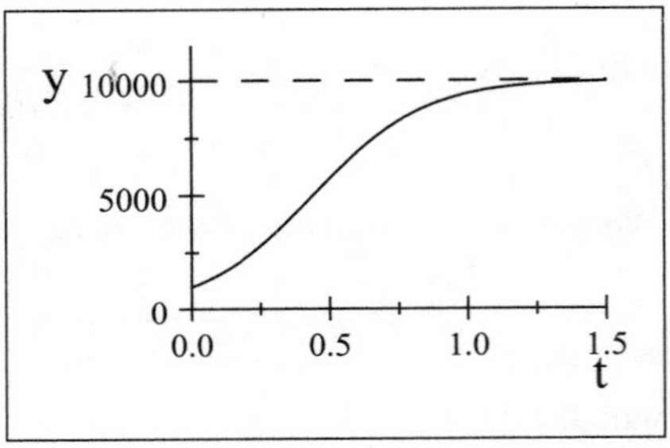

**Example 127** Suppose that a lake can support $C = 10,000$ fish and that the initial size of the fish population is $y(0) = 1000$. Assume that $t$ stands for time in years and $k = 0.0005$. To find the function $y(t)$ we set

$$y(0) = \frac{10,000}{1 + Ae^{0}} = 1000,$$

which leads to $A = 9$. We obtain

$$y(t) = \frac{10,000}{1 + 9e^{-5t}}.$$

The method we used to solve the logistic equation can be used for a wider class of differential equations. A **separable equation** has the form

$$\frac{dy}{dt} = f(t)\,g(y),$$

which means that the right-hand side is a product of two expressions: one depending only on $t$ and the other only on $y$.

**Example 128** Equations $y' = ye^{x}$, $y' = 3t$, and $y' = y(1 - y)$ are all separable, but equation $y' = t + y$ is not.

The first step in solving a separable equation is to divide both sides by $g(y)$, just as we did. Also, with a slight abuse of notation (although it can be formalized), we "multiply both sides by $dt$." Splitting the symbol $\frac{dy}{dt}$ can be avoided, as we just did, but it is convenient to allow this indulgence. (The discussion of the exact meaning of a stand-alone $dy$ or $dt$ is best avoided in polite society. Physicists like to call these expressions "infinitesimal changes of $y$ or $t$," but that is not too precise either. The good news is that this does not lead to any contradiction.) We write

$$\frac{1}{g(y)}dy = f(t)\,dt,$$

which is equivalent to $\frac{1}{g(y)}\frac{dy}{dt} = f(t)$. Integrating both sides with respect to $t$ we get

$$\int \frac{1}{g(y)}dy = \int f(t)\,dt.$$

As long as we can find both integrals, the problem is reduced to an algebraic equation.

**Example 129** Solve the initial value problem $y' + xy^2 = 0$, with $y(1) = 1$. First, we add $xy^2$ to both sides: $\frac{dy}{dx} = -xy^2$. Then we separate the variables and integrate:

$$\begin{aligned} \int -\frac{1}{y^2}dy &= \int x dx, \\ \frac{1}{y} &= \frac{1}{2}x^2 + C. \end{aligned}$$

To solve for $y$, we take the reciprocal of both sides:

$$y = \frac{1}{\frac{1}{2}x^2 + C} = \frac{2}{x^2 + C}.$$

To satisfy the initial condition we need $y(1) = \frac{2}{1+C} = 1$, which gives $C = 1$.

**WARNING:** For some reason, it is very common for students to write the reciprocal of $\frac{1}{2}x^2 + C$ as $\frac{2}{x^2} + C$, even though by now we certainly all know that $\frac{1}{a+b} \neq \frac{1}{a} + \frac{1}{b}$. Example 129 shows the importance of adding the constant of integration just after finding the antiderivative.

## 7.5 Problems

1. Find the length of the graph of $f(x) = \frac{1}{12}x^3 - \frac{1}{x}$ over the interval $[1, 2]$.

2. Let $F(x) = \int_0^x \sqrt{e^{2t} - 1}dt$. Find the arc length of the curve $y = F(x)$ over the interval $[0, 2]$.

3. Find the area bounded by the graphs of the equations: $y = \frac{1}{2}x^2$ and $y = \frac{1}{1+x^2}$.

4. Find the volume of a ball of radius $r$ using discs and using cylindrical shells.

5. A surface is created by rotating a graph of $y = f(x)$ about the interval $[a, b]$ on the $x$-axis. Set up the integral for the area of this surface. Then find the area of a sphere of radius $r$ by rotating the upper half of the circle $x^2 + y^2 = r^2$ about the $x$-axis.

6. Solve the initial value problem: $\frac{dy}{dt} = y^2 \sin t, \quad y(0) = \frac{1}{2}$.

7. (**Euler's method**) Consider an initial value problem $\frac{dy}{dt} = f(t, y), \; y(t_0) = y_0$. The values of $t_0$ and $y_0$ are given and $f(t, y)$ stands for some expression involving $t$ and $y$. Suppose that a function $y(t)$ solves the initial value problem.

   (a) Find an equation of the line tangent to the graph of $y(t)$ at the point $(t_0, y_0)$.

   (b) Use the equation of the tangent line to approximate the value of the function for $t_1 = t_0 + \Delta t$. Call this value $y_1$.

   (c) Repeat this procedure, but this time start with the point $(t_1, y_1)$.

   (d) Show that $y_{n+1} = y_n + f(t_n, y_n)\Delta t$.

   (e) Use this method to find a numerical approximation to the problem from Example 127.

# Chapter 8

# Infinite Sequences and Series

We have seen in Section 4.4 how to approximate functions by polynomials, sometimes with remarkable accuracy, at least on intervals of finite length. For example, using the Maclaurin polynomial for $f(x) = e^x$ we can say that

$$e^x \approxeq 1 + x + \frac{x^2}{2} + \frac{x^3}{3!} + \cdots + \frac{x^n}{n!} = \sum_{k=0}^{n} \frac{x^k}{k!}.$$

It would be nice to be able to replace the approximate equality "$\approxeq$" with the exact one "$=$," that is, to write

$$e^x = 1 + x + \frac{x^2}{2} + \frac{x^3}{3!} + \cdots + \frac{x^n}{n!} + \cdots = \sum_{k=0}^{\infty} \frac{x^k}{k!}. \tag{8.1}$$

This requires making some sense out of adding infinitely many numbers, which is denoted by the innocent looking "$+\cdots$". This is the goal of this chapter.

**Example 130** Assuming that the formula (8.1) is true and that it is legal to differentiate such an infinite "polynomial," let us find the derivative of $e^x$. As expected, we obtain

$$\begin{aligned}\frac{d}{dx}\left(e^{x}\right) &= 0+1+\frac{2x}{2}+\frac{3x^2}{3!}+\cdots+\frac{nx^{n-1}}{n!}+\cdots \\ &= 1+x+\frac{x^2}{2}+\frac{x^3}{3!}+\cdots+\frac{x^n}{n!}+\cdots=e^{x}.\end{aligned}$$

**Example 131** Let us replace the variable $x$ in formula (8.1) by $x^2$. We get

$$e^{x^2}=1+x^2+\frac{x^4}{2}+\frac{x^6}{3!}+\cdots+\frac{x^{2n}}{n!}+\cdots=\sum_{k=0}^{\infty}\frac{x^{2k}}{k!}.$$

Integrating term by term, we obtain

$$\begin{aligned}\int e^{x^2}dx &= \int\left(1+x^2+\frac{x^4}{2}+\frac{x^6}{3!}+\cdots+\frac{x^{2n}}{n!}+\cdots\right)dx \\ &= x+\frac{x^3}{3}+\frac{x^5}{5\cdot 2}+\frac{x^7}{7\cdot 3!}+\cdots+\frac{x^{2n+1}}{(2n+1)\,n!}+\cdots+C \\ &= \sum_{k=0}^{\infty}\frac{x^{2k+1}}{(2k+1)\,k!}+C.\end{aligned}$$

This should give us an incentive to study the fairly delicate concept of an infinite series and its convergence. There are many functions that are impossible to integrate in the traditional sense: we are unable to find neatly packaged antiderivatives. However, if under some conditions, we can differentiate and integrate such infinite sums term by term, then we can at least express certain integrals in the form of infinite sums.

## 8.1 Sequences and their limits

An **infinite sequence** (or just **sequence**, for short) is an infinite list of numbers of the general form

$$a_1, a_2, \ldots, a_k, a_{k+1}, \ldots.$$

Individual entries are called **terms** of the sequence. A standard notation for a sequence with the general term $a_k$ is $\{a_k\}_{k=1}^{\infty}$ (or simply $\{a_k\}$). Formally, a sequence is a real-valued function $a : \mathbb{N} \to \mathbb{R}$, whose domain consists of all positive integers. Traditionally, we write $a_n$ instead of $a(n)$.

We say that the sequence $\{a_k\}_{k=1}^{\infty}$ converges to a **limit** $L$, and write

$$\lim_{k\to\infty} a_k = L,$$

if $a_k$ approaches $L$ to within any desired tolerance as $k$ increases without bound.

Formally, for a finite $L$, we say that $\lim_{k\to\infty} a_k = L$ if for every $\varepsilon > 0$, there exists an $N > 0$ such that $|a_k - L| < \varepsilon$ whenever $k \geq N$. We say that $\lim_{k\to\infty} a_k = \infty$ if for every $M > 0$, there exists an $N > 0$ such that $a_k > M$ whenever $k \geq N$.

## 8.2 Infinite series

We would like to express functions as infinite "polynomials," instead of just approximating them. The goal of this section is to make sense out of expressions involving adding infinitely many numbers, such as

$$\sum_{k=1}^{\infty} a_k = a_1 + a_2 + a_3 + \cdots + a_k + a_{k+1} + \cdots. \tag{8.2}$$

We will use the word **series** or **infinite series** to refer to an expression of this form. What could these mysterious dots at the end of the sum mean?

Given a sequence $\{a_k\}_{k=1}^{\infty}$ we define another sequence $\{S_n\}_{n=1}^{\infty}$, called the **sequence of partial sums** (of the original sequence):

$$S_n = \sum_{k=1}^{n} a_k = a_1 + a_2 + a_3 + \cdots + a_n.$$

It is sensible to say that the larger the value of $n$ we take, the closer the $n$-th partial sum $S_n$ is to the sum of all the terms.

Formally, if the sequence $S_n$ converges to $S$, we say that the series $\sum_{k=1}^{\infty} a_k$ **converges** and we call $S$ the **sum** of the series $\sum_{k=1}^{\infty} a_k$, that is,

$$\sum_{k=1}^{\infty} a_k = \lim_{n\to\infty} S_n = S.$$

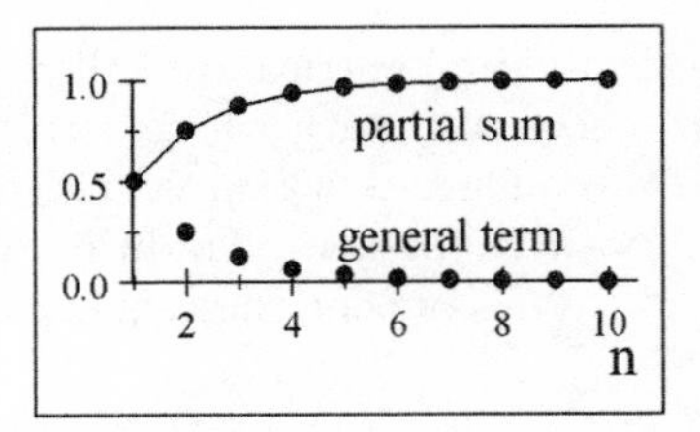


$a_n = 1/2^n \quad S_n = \sum_{k=1}^{n} \frac{1}{2^k}$

Otherwise we say that the series **diverges**. Frequently, we speak about the infinite series $\sum a_k$ without even specifying the exact range of values of $k$. It is good to think about a series $\sum a_k$ as an object involving two sequences: the sequence $\{a_k\}$ of **terms** and the sequence $\{S_n\}$ of **partial sums**.

**WARNING:** It is very common for people to use the phrase "it converges" while discussing infinite series. However, one should be careful as to what is "it": the general term or the series.

**Example 132** Does the series $\sum_{k=1}^{\infty} \frac{1}{2^k} = \frac{1}{2} + \frac{1}{4} + \frac{1}{8} + \cdots$ converge? The general term $a_k = \frac{1}{2^k}$ for all integer values of $k$. The sequence $\{a_k\}$ converges to 0, but that's not the question. The partial sums are

$$\begin{aligned} S_1 &= \frac{1}{2} = 1 - \frac{1}{2}, \\ S_2 &= \frac{1}{2} + \frac{1}{4} = \frac{3}{4} = 1 - \frac{1}{4}, \\ S_3 &= \frac{1}{2} + \frac{1}{4} + \frac{1}{8} = 1 - \frac{1}{8}, \\ &\vdots \\ S_n &= \sum_{k=1}^{n} \frac{1}{2^k} = 1 - \frac{1}{2^n}; \end{aligned}$$

therefore

$$S = \lim_{n\to\infty} S_n = \lim_{n\to\infty} \left(1 - \frac{1}{2^n}\right) = 1,$$

which means that the series $\sum_{k=1}^{\infty} \frac{1}{2^k}$ converges to 1. For short, we usually just write $\sum_{k=1}^{\infty} \frac{1}{2^k} = 1$.

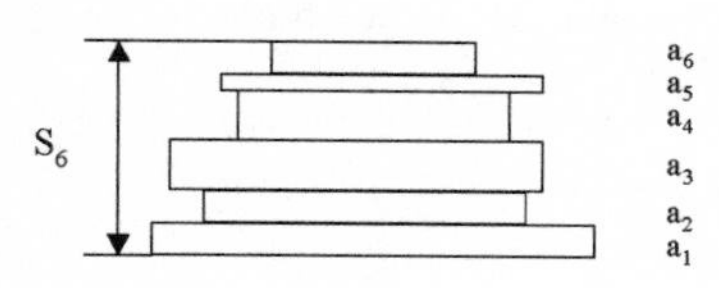


Let us assume for the moment that all the terms $a_k$ are positive. A good way to visualize the concept of the series is to imagine a stack made of infinitely many bricks of height $a_k$. (Think of a stack of books on your favorite professor's desk.) Clearly, if there is any hope that this stack has finite height, the terms $a_k$ must be getting smaller and smaller. That is, in order for the infinite sum $\sum_{k=1}^{\infty} a_k = \lim S_n$ to converge to *something,* it is necessary that $\lim_{k\to\infty} a_k = 0$. Another way to see that is to realize that $a_k = S_k - S_{k-1}$, for all $k$ (or at least for $k \geq 2$). If $\lim_{k\to\infty} S_k = S$ exists (a finite limit, that is), then for sufficiently large $k$, the values of $S_k$ are close to $S$. Therefore, the difference $S_k - S_{k-1}$ must be small.

**Example 133** Consider $\sum_{k=1}^{\infty} \frac{k}{k+1}$. The sequence $a_k = \frac{k}{k+1}$ converges to 1 as $k \to \infty$. Since 1 is not 0, the series diverges.

Once again, we have established that in order for the series $\sum_{k=1}^{\infty} a_k$ to converge (to something), the general term $a_k$ must converge to 0. It is important to realize that this *necessary condition* for the convergence of a series is *not sufficient.* The following example shows that the convergence of a series is not a trivial matter.

**Example 134** The **harmonic series** $\sum_{k=1}^{\infty} \frac{1}{k} = \frac{1}{1} + \frac{1}{2} + \frac{1}{3} + \cdots$ diverges. Let us first notice that

$$\begin{aligned}
\frac{1}{3} + \frac{1}{4} &> \frac{1}{4} + \frac{1}{4} = \frac{1}{2}, \\
\frac{1}{5} + \frac{1}{6} + \frac{1}{7} + \frac{1}{8} &> \frac{1}{8} + \frac{1}{8} + \frac{1}{8} + \frac{1}{8} = \frac{1}{2}, \\
\frac{1}{9} + \frac{1}{10} + \frac{1}{11} + \frac{1}{12} + \frac{1}{13} + \frac{1}{14} + \frac{1}{15} + \frac{1}{16} &> 8 \cdot \frac{1}{16} = \frac{1}{2}, \\
&\vdots \\
\frac{1}{2^i + 1} + \frac{1}{2^i + 2} + \frac{1}{2^i + 3} + \cdots + \frac{1}{2^{i+1}} &> 2^i \cdot \frac{1}{2^{i+1}} = \frac{1}{2}.
\end{aligned}$$

Hence, by taking sufficiently many terms of the harmonic series we have

$$\begin{aligned} S_{2^{i+1}} &= 1 + \frac{1}{2} + \left(\frac{1}{3} + \frac{1}{4}\right) + \left(\frac{1}{5} + \frac{1}{6} + \frac{1}{7} + \frac{1}{8}\right) + \cdots \\ &\quad + \left(\frac{1}{2^i+1} + \frac{1}{2^i+2} + \frac{1}{2^i+3} + \cdots + \frac{1}{2^{i+1}-1} + \frac{1}{2^{i+1}}\right) \\ &\quad > 1 + \frac{1}{2} + \frac{1}{2} + \frac{1}{2} + \cdots + \frac{1}{2} = 1 + \frac{i+1}{2} \end{aligned}$$

(where $\frac{1}{2}$ in the sum above is repeated $i+1$ times). For instance, adding the first $2^{10} = 1024$ terms of the series gives $S_{1024} > 1 + \frac{10}{2} = 6$. In the same way: $S_{2^{20}} = S_{1,048,576} > 1 + \frac{20}{2} = 11$, and so on. The partial sums grow very slowly, but we still have

$$\sum_{k=1}^{\infty} \frac{1}{k} = \lim_{n\to\infty} S_n = \infty,$$

which means that the harmonic series diverges to $\infty$.

Example 134 shows that just because the general term $a_k$ converges to 0, we cannot claim that the series $\sum a_k$ converges to anything. For the series $\sum a_k$ to converge, the general term $a_k$ must approach 0 *sufficiently fast.*

Example 132 is a special case of an important class of infinite series called geometric. Partial sums of a geometric series can be expressed in a nice, short form, which is a rare and pleasant occurrence for a series. In general, a **geometric sequence** $\{a_k\}$ is defined by the property that the ratio of two successive terms $a_{k+1}/a_k = r$ is a constant. In Example 132 this constant was $\frac{1}{2}$. In other words, a geometric sequence is given by the formula $a_k = ar^k$, where $a$ is the first term of the sequence (corresponding to $k = 0$) and $r$ is the quotient of every two subsequent terms. We now consider a **geometric series** $\sum_{k=0}^{\infty} ar^k$. (Traditionally, a geometric series starts with $k = 0$, although this is not essential.) Consider a partial sum

$$S_n = \sum_{k=0}^{n} ar^k = a + ar + ar^2 + \cdots + ar^n. \tag{8.3}$$

It is a simple, but neat, trick to multiply both sides of (8.3) by the expression $1-r$. We get

$$\begin{aligned} S_n(1-r) &= \left(a+ar+ar^2+\cdots ar^n\right)(1-r) \\ &= a-ar+ar-ar^2+\cdots+ar^n-ar^{n+1} \\ &= a-ar^{n+1}=a\left(1-r^{n+1}\right), \end{aligned}$$

since all the terms cancel except for the first and the last one. Dividing by $1-r$ we obtain

$$S_n = a\frac{1-r^{n+1}}{1-r}. \tag{8.4}$$

It is easy to see that the behavior of the $r^{n+1}$ term for $n\to\infty$ in formula (8.4) depends on the value of $r$. If $|r|<1$, then $\lim_{n\to\infty} r^{n+1}=0$ (a fraction raised to a large power is close to 0). If $r=1$, then formula (8.4) cannot be used, but the series clearly diverges, as $S_n=a+a+\cdots+a=(n+1)a$. For $r>1$, the expression $r^{n+1}\to\infty$ as $n\to\infty$. Finally, for $r\le -1$, $r^{n+1}$ oscillates in sign and gets large in size (unless $r=1$) with $n\to\infty$.

We have established that the geometric series converges only for $-1<r<1$, and

$$\sum_{k=0}^{\infty} ar^k = \frac{a}{1-r}. \tag{8.5}$$

For $r\ge 1$, the series diverges to $\pm\infty$, depending on the sign of $a$; for $r\le -1$, the series just diverges (to nothing in particular). The fraction in formula (8.5) can be remembered as the quotient of the first term over $1-r$. A nice thing about this wording is that the labeling of the sequence does not matter: we can start with $k=0$ or with any other integer value.

**Example 135** Let us take another look at Example 132. In this case we can think of $\sum_{k=1}^{\infty}\frac{1}{2^k}=\frac{1}{2}+\frac{1}{4}+\frac{1}{8}+\cdots$ as $\sum_{i=0}^{\infty} ar^i$, where $a=\frac{1}{2}$ and $r=\frac{1}{2}$. Indeed, for $i=0$ in the second series we get the same number as for $k=1$ in the first one: $\frac{1}{2}$. All we did is shift the label by one: $i=k-1$. The sum of the infinite series, by formula (8.5), is $\sum_{i=0}^{\infty} ar^i=\sum_{i=0}^{\infty}\frac{1}{2}\left(\frac{1}{2}\right)^i=\frac{a}{1-r}=\frac{1/2}{1-1/2}=1$, as we saw in Example 132.

**Example 136** Consider the series: $5 + 10 + 20 + 40 + \cdots = \sum_{k=0}^{\infty} 5 \cdot 2^k$, which obviously diverges. This is confirmed by formula (8.4): the ratio $r = 2$ and the $n$-th partial sum $S_n = \sum_{k=0}^{n} 5 \cdot 2^k = 5\frac{1-2^{n+1}}{1-2} = 5\left(2^{n+1} - 1\right)$. Clearly, $\lim_{n\to\infty} S_n = \infty$.

## 8.3 Convergence tests for series

In this section, we consider *only nonnegative* series, that is, the general term $a_k \geq 0$, for all $k$. This excludes the possibility that the sequence of partial sums has no limit. When $a_k \geq 0$, then the sequence $S_n$ is increasing (nondecreasing to be exact) and therefore it either converges to a finite number or $\lim_{n\to\infty} S_n = \infty$. This property of all monotone sequences is intuitively clear, but a rigorous proof relies on properties of the real numbers which are beyond the scope of this book. The convergence tests discussed in this section help us discern whether the partial sums are bounded and the series converges, or increase without bound and the series diverges.

### 8.3.1 Comparison test

The main idea is to replace a more complicated series with a simpler one whose convergence we can determine. However, one has to be careful not to jump to the wrong conclusions. To show convergence of a given series, we can find a friendly convergent series greater than the one in question. In the opposite direction, to establish divergence of a given series, one should find a smaller series which diverges. It is useful to develop some intuition about the behavior of a given series before attempting to use the comparison test. More formally, suppose that $0 \leq a_k \leq b_k$, say, for all $k \geq 0$. If $\sum b_k$ converges, then $\sum a_k$ converges and $\sum_{k=0}^{\infty} a_k \leq \sum_{k=0}^{\infty} b_k$; if $\sum a_k$ diverges, then $\sum b_k$ diverges.

In the statement above, as well as in most statements regarding convergence of infinite series, it is not essential that something is true for all $k$, just that it happens eventually, that is, for all sufficiently large $k$, say $k > N$, where $N$ is some large number. (What happens for the first million terms does not matter.)

**Example 137** Consider $\sum_{k=2}^{\infty} \frac{2^k}{3^k+5}$. We do not know what the series converges to, but we know that it converges because $\frac{2^k}{3^k+5} < \frac{2^k}{3^k}$. Therefore $\sum_{k=2}^{\infty} \frac{2^k}{3^k+5} < \sum_{k=2}^{\infty} \frac{2^k}{3^k} = \frac{2^2}{3^2} \cdot \frac{1}{1-2/3} = \frac{4}{3}$ (as a sum of the geometric series with $r = \frac{2}{3}$ and the first term $\frac{2^2}{3^2}$).

**Example 138** The series $\sum_{k=0}^{\infty} \frac{1}{\sqrt{k}+6}$ diverges, because $\frac{1}{\sqrt{k}+6} > \frac{1}{k}$, for all $k > 9$ and the harmonic series $\sum \frac{1}{k}$ diverges.

### 8.3.2 Integral test

The series $\sum a_k$ and integrals of the form $\int_1^{\infty} a(x)\,dx$ are closely related. Precisely, suppose that $a_k = a(x)$ for some *positive, continuous, decreasing* function $a(x)$. Then the series $\sum a_k$ and the integral $\int_1^{\infty} a(x)\,dx$ **both converge** or **both diverge**. This is obvious after looking at the following pictures:

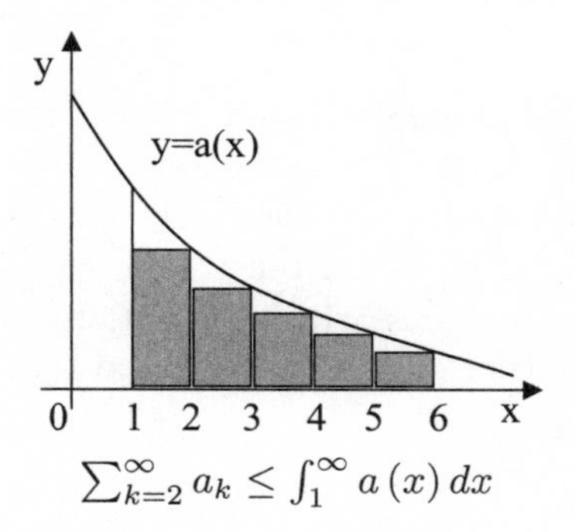


$\sum_{k=2}^{\infty} a_k \leq \int_1^{\infty} a(x)\,dx$

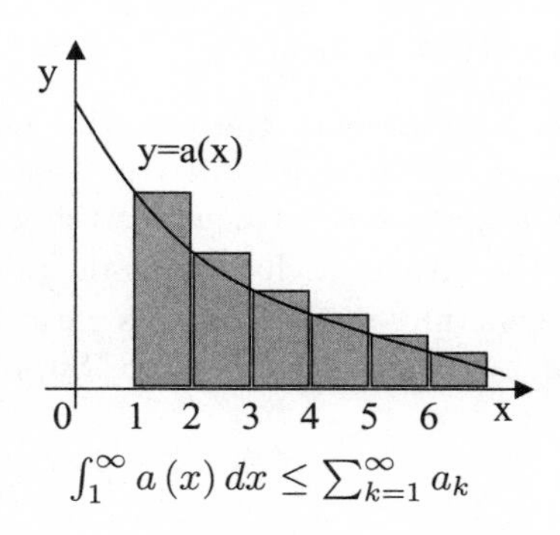


$\int_1^{\infty} a(x)\,dx \leq \sum_{k=1}^{\infty} a_k$

In the first graph, the heights of the rectangles correspond to the values $a_2$, $a_3, \ldots$. The rectangles all fit under the graph as they are drawn "to the left" of the corresponding values of $n$. Similarly, the rectangles in the second graph all stick out above the graph of $a(x)$. Therefore

$$\sum_{k=2}^{\infty} a_k \leq \int_1^{\infty} a(x)\,dx \leq \sum_{k=1}^{\infty} a_k.$$

Since $\sum_{k=1}^{\infty} a_k = a_1 + \sum_{k=2}^{\infty} a_k$, we also have

$$\int_1^{\infty} a(x)\,dx \leq \sum_{k=1}^{\infty} a_k \leq a_1 + \int_1^{\infty} a(x)\,dx.$$

**Example 139** We already know that the harmonic series from Example 134 diverges. Another way to see it is to apply the integral test: the function $a(x) = \frac{1}{x}$ is decreasing and $a(n) = \frac{1}{n}$. As expected, we have $\int_1^{\infty} \frac{1}{x} dx = \lim_{t\to\infty} \int_1^t \frac{1}{x} dx = \lim_{t\to\infty} (\ln t - \ln 1) = \infty$.

**Example 140** For which values of $p$ does the series $\sum \frac{1}{k^p}$ converge? Consider a function $a(x) = \frac{1}{x^p}$. We know from Section 5.5 that $\int_1^{\infty} \frac{1}{x^p} dx$ converges if and only if $p > 1$. Therefore, the same is true for the series in question. The series $\sum \frac{1}{k^p}$ is called the $p$-series, and it is one of the "friendly series" frequently used in the comparison test.

### 8.3.3 Ratio test

We are familiar with the behavior of the geometric series $\sum ar^k$: the series converges whenever $|r| < 1$ (see formula (8.5)). The ratio test measures how closely a given series resembles the geometric series. For the geometric series $\sum ar^k$ the ratio of each term to the previous one is constant: $\frac{a_{k+1}}{a_k} = \frac{ar^{k+1}}{ar^k} = r$. For a general series $\sum a_k$ this ratio is not necessarily constant, but it may approach a constant as $k \to \infty$. Suppose that $a_k > 0$ for all $k$ and that

$$\lim_{k\to\infty} \frac{a_{k+1}}{a_k} = L.$$

- If $L < 1$, then $\sum a_k$ converges.
- If $L > 1$, then $\sum a_k$ diverges.
- If $L = 1$, then the ratio test is inconclusive.

(The proof relies on comparing the series with the geometric series $\sum aL^k$.) If some of the terms of a series are negative, we can still use the ratio test, but we calculate $L = \lim_{k\to\infty} \left|\frac{a_{k+1}}{a_k}\right|$. The ratio test works well for many series in which the index $k$ appears in an exponent or with a factorial. Unfortunately, for many series the limit $L = 1$ or does not exist at all. If that happens, we need to try another method.

**Example 141** Let us try the ratio test on a series $\sum_{k=2}^{\infty} \frac{2^k}{3^k+5}$ from Example 137. We have $a_k = \frac{2^k}{3^k+5}$ and $a_{k+1} = \frac{2^{k+1}}{3^{k+1}+5}$. We leave it to the reader to verify that the ratio $\frac{a_{k+1}}{a_k} = \frac{2^{k+1}}{3^{k+1}+5} \cdot \frac{3^k+5}{2^k}$ converges to $L = \frac{2}{3}$ as $k \to \infty$ (one can divide the numerators and denominators by $2^k$ and by $3^{k+1}$ accordingly, or apply l'Hôpital's rule to $\frac{3^x+5}{3^{x+1}+5}$). Since $\frac{2}{3} < 1$, the series converges.

**Example 142** We already know from Example 140 that the series $\sum \frac{1}{k^p}$ converges for $p < 1$. However, an attempt to use the ratio test to establish this fact is futile. We have $\frac{a_{k+1}}{a_k} = \frac{k^p}{(k+1)^p} = \left(\frac{k}{k+1}\right)^p \to 1$ since $\frac{k}{k+1} \to 1$ as $k \to \infty$. The ratio test is inconclusive.

**Example 143** Let us consider the series $\sum_{k=0}^{\infty} \frac{1}{k!}$ (where $k! = 1 \cdot 2 \cdot 3 \cdot \cdots \cdot k$ for $k \geq 1$, and $0! = 1$, by definition). The ratio test gives $\frac{a_{k+1}}{a_k} = \frac{k!}{(k+1)!} = \frac{k!}{(k+1)\cdot k!} = \frac{1}{k+1} \to 0$ as $k \to \infty$. Since $0 < 1$, the series converges. By the way, $\sum_{k=0}^{\infty} \frac{1}{k!} = e$, as we will see in Section 8.5.

## 8.4 Absolute convergence, alternating series

We now return to considering series whose general term $a_k$ *may change sign.* Some new terminology is needed. Let $\sum a_k$ be any series. If $\sum |a_k|$ converges, then we say that $\sum a_k$ **converges absolutely**. If $\sum |a_k|$ diverges and $\sum a_k$ converges, then we say that $\sum a_k$ **converges conditionally**.

This definition only makes sense if the convergence of $\sum |a_k|$ implies the convergence of $\sum a_k$. Luckily, this is the case. Moreover, $|\sum_{k=1}^{\infty} a_k| \leq \sum_{k=1}^{\infty} |a_k|$. We summarize the possibilities in a table:

| | $\sum a_k$ converges | $\sum a_k$ diverges |
|---|---|---|
| $\sum \|a_k\|$ converges | $\sum a_k$ converges absolutely | not possible |
| $\sum \|a_k\|$ diverges | $\sum a_k$ converges conditionally | $\sum a_k$ diverges |

**Example 144** The series

$$\sum_{k=1}^{\infty} \frac{(-1)^{k+1}}{k} = 1 - \frac{1}{2} + \frac{1}{3} - \frac{1}{4} + \frac{1}{5} + \cdots$$

converges (we will see why in a moment), while the harmonic series

$$\sum_{k=1}^{\infty} \frac{1}{k} = 1 + \frac{1}{2} + \frac{1}{3} + \frac{1}{4} + \frac{1}{5} + \cdots$$

diverges. Hence the series $\sum_{k=1}^{\infty} \frac{(-1)^{k+1}}{k}$ converges conditionally.

**Example 145** The series $\sum_{k=1}^{\infty} \frac{\sin k}{k^2}$ converges absolutely. This is because the series $\sum_{k=1}^{\infty} \frac{|\sin k|}{k^2}$ converges. We know that by comparing this series with a friendly one, which converges: $\frac{|\sin k|}{k^2} \leq \frac{1}{k^2}$.

In mathematical analysis absolute convergence is much preferred over conditional. The name "conditional" indicates that the convergence depends upon something. Absolutely convergent series have the nice property that the sum does not depend on the order of terms, that, is that addition remains commutative in such infinite sums. However, this is not true for conditionally convergent series, for which the sum depends on the order of terms.

In fact, for any conditionally convergent series and any real number $S$, one can rearrange the terms of the series so that the new series converges to $S$. One can also rearrange them so that the new series diverges. This is perhaps counterintuitive, as we expect addition always to be commutative. However, when infinitely many terms are reshuffled, our intuition may fail. An inquisitive reader is probably curious at this point how can this be done. It goes more or less like this. Imagine that we would like the new series to converge to 5. First, we add as many positive terms as needed, until we get just above 5. Next, we start using negative terms until we get below 5. Then we use positive terms, until we get just above 5 again, and so on.

**Alternating series test:** *Consider the series*

$$\sum_{k=1}^{\infty} (-1)^{k+1} c_k = c_1 - c_2 + c_3 - c_4 + \cdots, \tag{8.6}$$

*where*

$$c_1 \geq c_2 \geq c_3 \geq \cdots \geq 0 \qquad \text{and} \lim_{k\to\infty} c_k = 0. \tag{8.7}$$

*Then the series converges and its sum $S$ lies between any two successive partial sums $S_n$ and $S_{n+1}$. In particular,*

$$|S - S_n| < c_{n+1}$$

*for all $n \geq 1$.*

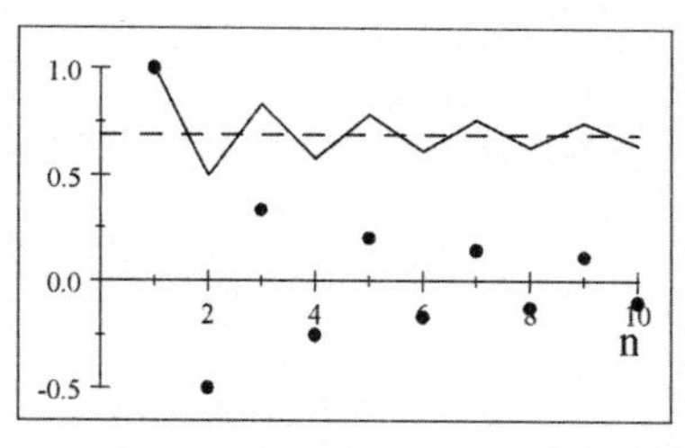


$a_n = \frac{(-1)^n}{n} \quad S_n = \sum_{k=1}^{n} \frac{(-1)^k}{k}$

This is one of those rare occasions when dancing helps us understand calculus. Let us consider adding a positive term as making a step to the right; adding a negative term corresponds to moving to the left. The condition (8.7) means that the steps are getting smaller and eventually we hardly move at all. This explains why the partial sums $S_n$ have a limit $S$. Also, since the steps are getting smaller and we switch direction at every step, the $n$-th partial sum is closer to $S$ than the size of the next step $c_{n+1}$.

**Example 146** The series $\sum_{k=1}^{\infty} \frac{(-1)^{k+1}}{\sqrt{k}}$ converges conditionally. To see that the original series converges, we apply the alternating series test. In our case $c_k = \frac{1}{\sqrt{k}}$ and it is clear that $c_{k+1} < c_k$ and that $c_k \to 0$ as $k \to \infty$, so the series converges, On the other hand, the series $\sum_{k=1}^{\infty} \left|\frac{(-1)^{k+1}}{\sqrt{k}}\right| = \sum_{k=1}^{\infty} \frac{1}{\sqrt{k}}$ diverges as a $p$-series with $p = \frac{1}{2}$. Hence, the convergence of the original series is conditional and not absolute. Using more or less the same argument we can see that the alternating harmonic series from Example 144 is also conditionally convergent.

**Example 147** To determine the type of convergence of the series $\sum_{k=1}^{\infty} \frac{(-1)^{k+1}}{2^k+7}$ we notice that the series of absolute values $\sum_{k=1}^{\infty} \frac{1}{2^k+7}$ converges, by comparison with the geometric series $\sum_{k=1}^{\infty} \frac{1}{2^k}$. Hence, the original series converges absolutely. There is no need to bother with the alternating series test.

## 8.5 Power series as functions, Taylor series

We started this chapter with the series (8.1): $\sum_{k=0}^{\infty} \frac{1}{k!} x^k$. We are now ready to take a closer look. The first question is whether the series converges. Since we now have a variable as a part of the general term, convergence may depend on the value of $x$. The sum of the series is also a function of $x$. The domain of this function consists of those values of $x$ for which the series converges. In this section, we consider a class of series whose general term depends on $x$ in a very specific way. Series of the form $\sum_{k=0}^{\infty} a_k (x - x_0)^k$ are called **Power series.** The constants $a_k$ are called the **coefficients,** and the constant $x_0$ is called the **base point**.

**Example 148** Let us use the ratio test to determine for what values of $x$ the series $\sum_{k=0}^{\infty} \frac{1}{k!} x^k$ converges. We have

$$\lim_{k\to\infty} \frac{\left|\frac{1}{(k+1)!} x^{k+1}\right|}{\left|\frac{1}{k!} x^k\right|} = \lim_{k\to\infty} \left|\frac{k!}{(k+1)!} \frac{x^{k+1}}{x^k}\right|$$
$$= \lim_{k\to\infty} \left|\frac{k!}{(k+1)\cdot k!}\right| |x| = \lim_{k\to\infty} \left|\frac{1}{k+1}\right| |x| = 0.$$

The limit turns out to be 0, regardless of the value of $x$. Since $0 < 1$, the series converges for all values of $x$.

**Example 149** Consider $\sum_{k=0}^{\infty} \frac{(x+1)^k}{3^k}$. This series is geometric, so using the ratio test is an overkill. Still, we get:

$$\lim_{k\to\infty} \frac{\left|\frac{(\mathbf{x}+\mathbf{1})^{k+1}}{3^{k+1}}\right|}{\left|\frac{(\mathbf{x}+\mathbf{1})^k}{3^k}\right|} = \lim_{k\to\infty} \left|\frac{3^k}{3^{k+1}} \frac{(\mathbf{x}+\mathbf{1})^{k+1}}{(\mathbf{x}+\mathbf{1})^k}\right| = \lim_{k\to\infty} \left|\frac{1}{3}(\mathbf{x}+\mathbf{1})\right| = \frac{1}{3}|\mathbf{x}+\mathbf{1}|.$$

For the series to converge, we need this limit $\frac{1}{3}|x+1| < 1$, which means $|x+1| < 3$. The series is convergent for $x \in (-4, 2)$ and it is divergent for $x \in (-\infty, -4) \cup (2, \infty)$. This is all we can say based on the ratio test. However, for $x = 2$ we have $\sum_{k=0}^{\infty} \frac{(2+1)^k}{3^k} = \sum_{k=0}^{\infty} \frac{3^k}{3^k} = \sum_{k=0}^{\infty} 1 = \infty$. For $x = -4$, things are not much better: $\sum_{k=0}^{\infty} \frac{(-4+1)^k}{3^k} = \sum_{k=0}^{\infty} \frac{(-3)^k}{3^k} = \sum_{k=0}^{\infty} (-1)^k$, which diverges.

The series $\sum_{k=0}^{\infty}(-1)^k = 1 - 1 + 1 - 1 + 1 - 1 + \cdots$ is obviously divergent. However, it does make some sense to assciate a number $\frac{1}{2}$ with this series. After all, $S_0$=1, $S_1$=0, $S_2$=1, $S_3$=0, and so on. The "average" value of $S_n$ should be $\frac{1}{2}$. Various ways of assigning such "sums" to divergent series are known as summability methods.

The set of those values of $x$ for which a power series converges is always a symmetric interval centered at $x_0$. We leave this statement without a formal proof, but the general idea is that if a power series converges absolutely for some value of $x$, then it must also converge absolutely for every value which is closer to the base $x_0$. This can be verified using the comparison test. There is always a single value $R$ (which can sometimes be 0 or $\infty$), called the **radius of convergence**, such that the power series converges absolutely whenever the distance from the base $x_0$ to $x$, $|x - x_0| < R$. The series diverges if $|x - x_0| > R$. What happens on each of the border points, that is, at $x = x_0 \pm R$, depends on the series. The behavior of a power series at the edge of the region of convergence should examined separately, if necessary, as the series might converge absolutely, conditionally or diverge. The radius of convergence is usually found using the ratio test.

Yes, it is a little unusual to describe an interval by its center and radius. This terminology has roots in complex analysis: the region of convergence of any power series is a disk on the complex plane, centered at base point $x_0$.

The following fact is very useful, as integrating polynomials is easy and we are now able to find some formulas for antiderivatives of functions previously impossible to integrate, albeit in the form of a series. The proof of the following theorem is too advanced for the level of this book. The reader is asked to make a leap of faith. (Perhaps I have earned the reader's trust by now?)

**Theorem:** *A function written as a power series can be differentiated and integrated "term by term" for $x$ strictly inside the region of convergence. Formally, suppose*

$$f(x) = \sum_{k=0}^{\infty} a_k (x - x_0)^k, \tag{8.8}$$

*with the radius of convergence $R > 0$. For every $x \in (x_0 - R, x_0 + R)$ the function is differentiable and its derivative and antiderivative can also be written* *as a power series with the same radius of convergence R:*

$$f'(x) = \sum_{k=1}^{\infty} k a_k (x - x_0)^{k-1},$$

$$\int f(x)\,dx = C + \sum_{k=0}^{\infty} \frac{a_k}{k+1}(x - x_0)^{k+1}.$$

There is of course absolutely no need to memorize the formulas for $f'(x)$ and $\int f(x)\,dx$, as they are obtained by differentiating (or integrating) term by term. The following examples illustrate the use of the theorem.

**Example 150** The power series $\sum_{n=0}^{\infty} x^n$ is actually geometric. Not only do we know that it converges for $x \in (-1,1)$, we even know that

$$f(x) = \sum_{n=0}^{\infty} x^n = \frac{1}{1-x}. \tag{8.9}$$

Let us integrate the series term by term:

$$\int f(x)\,dx = C + x + \frac{x^2}{2} + \frac{x^3}{3} + \frac{x^4}{4} + \cdots = C + \sum_{k=1}^{\infty} \frac{x^k}{k},$$

and since $\int f(x)\,dx = -\ln(1-x) + \overline{C}$ we obtain (possibly adjusting the constant $C$)

$$-\ln(1-x) = C + \sum_{k=1}^{\infty} \frac{x^k}{k}.$$

Plugging $x = 0$ into both sides gives $-\ln 1 = C + 0$, hence $C = 0$. Therefore

$$\ln(1-x) = -\sum_{k=1}^{\infty} \frac{x^k}{k},$$

for all $x \in (-1,1)$. Replacing $1-x$ with $t$, we can write

$$\ln t = -\sum_{k=1}^{\infty} \frac{(1-t)^k}{k} = \sum_{k=1}^{\infty} \frac{(-1)^{k+1}}{k}(t-1)^k$$

for all $t \in (0,2)$.

**Example 151** This time let us differentiate the series (8.9):

$$f'(x) = \frac{1}{(1-x)^2} = 0 + 1 + 2x + 3x^2 + \cdots = \sum_{n=1}^{\infty} nx^{n-1},$$

which can also be written as $\sum_{k=0}^{\infty} (k+1) x^k$.

Suppose that a function has a power series representation (8.8) on an interval $(x_0 - R, x_0 + R)$. It is good to know that this representation is unique, meaning that there is only one sequence of coefficients $\{a_n\}$ which makes (8.8) true. This is not hard to check. Writing it all out for convenience and differentiating both sides we get

$$\begin{aligned}
f(x) &= a_0 + a_1 (x - x_0) + a_2 (x - x_0)^2 + a_3 (x - x_0)^3 + \cdots \\
f'(x) &= a_1 + 2a_2 (x - x_0) + 3a_3 (x - x_0)^2 + 4a_4 (x - x_0)^3 + \cdots \\
f''(x) &= 2a_2 + 3 \cdot 2a_3 (x - x_0) + 4 \cdot 3a_4 (x - x_0)^2 + 5 \cdot 4a_5 (x - x_0)^3 + \cdots \\
&\vdots \\
f^{(n)}(x) &= n!a_n + ((n+1) n (n-1) \cdot \cdots \cdot 2) a_{n+1} (x - x_0) + \cdots
\end{aligned}$$

and after we put $x = x_0$ we obtain

$$f(x_0) = a_0, \quad f'(x_0) = a_1, \quad f''(x_0) = 2a_2, \quad \ldots \quad f^{(n)}(x_0) = n!a_n, \quad \ldots$$

which shows that the coefficients $\{a_n\}$ are unique and in fact they are precisely the Taylor coefficients

$$a_n = \frac{f^{(n)}(x_0)}{n!}.$$

Wrapping it up, we have established that if a function has a convergent power series (8.8) on an interval $(x_0 - R, x_0 + R)$ then

$$f(x) = \sum_{k=0}^{\infty} \frac{f^{(n)}(x_0)}{n!} (x - x_0)^k. \tag{8.10}$$

There is still the issue of whether the Taylor series converges and on what interval. Using the bounds for the Taylor polynomial from Section 4.4 one can fairly easily establish that formula (8.10) is valid as long as there is a single constant $K$, such that for all values of $k \geq 0$ we have $\left|f^{(k)}(x_0)\right| \leq K$ for all $x \in (x_0 - R, x_0 + R)$.

We can now joyfully pronounce that for all $x \in \mathbb{R}$

$$\begin{aligned} e^x &= 1 + x + \frac{x^2}{2!} + \frac{x^3}{3!} + \cdots = \sum_{n=0}^{\infty} \frac{x^n}{n!}, \\ \sin x &= x - \frac{x^3}{3!} + \frac{x^5}{5!} - \frac{x^7}{7!} + \cdots = \sum_{n=0}^{\infty} \frac{(-1)^n x^{2n+1}}{(2n+1)!}, \\ \cos x &= 1 - \frac{x^2}{2!} + \frac{x^4}{4!} - \frac{x^6}{6!} + \cdots = \sum_{n=0}^{\infty} \frac{(-1)^n x^{2n}}{(2n)!}. \end{aligned} \tag{8.11}$$

For every function which is repeatedly differentiable at $x_0$ one can construct the Taylor series. Some care is needed regarding the exact region where the series actually converges to the function we started with. The series may have a small interval of convergence (or even 0 if we are very unlucky). In addition, it may converge to something other than $f(x)$. Luckily, this does not happen often.

## 8.6 Invitation to harmonic analysis

The reader is asked for a small indulgence at this point, as we are about to make a little side trip into the world of infinite-dimensional spaces. Hope you enjoy the ride!

There are other "nice families" of functions which could be used instead of polynomials. We mention in passing, for brevity's sake, that one such family consists of trigonometric polynomials, that is, functions built as linear combinations of the basic trigonometric functions:

$$q_n(x) = a_0 + a_1 \cos x + b_1 \sin x + \cdots + a_n \cos(nx) + b_n \sin(nx).$$

It turns out that trigonometric polynomials approximate periodic functions much better than regular polynomials.

There is a very elegant way of looking at functions as vectors in a suitable infinite-dimensional vector space. One can generalize the "dot product" in a 3-dimensional space to this much more spacious vectors space of functions. The generalized dot product is usually called the **inner product**. Just as two vectors in space are perpendicular, or orthogonal, whenever their dot product is zero, any two functions are defined to be orthogonal whenever their inner product is 0. For example, for continuous functions of period $2\pi$, we can define the inner product of two functions $\langle f, g\rangle = \int_{-\pi}^{\pi} f(x)\, g(x)\, dx$. The trigonometric polynomials form an orthogonal family of functions.

The idea of the best approximation with respect to this inner product is very geometrical in nature and is due to Fourier. It is actually quite beautiful in its simplicity. Imagine a point and a plane in a good old-fashioned 3-dimensional Euclidean space $\mathbb{R}^3$. The best approximation of any given point by another point from the given plane is such that the line connecting the two points is perpendicular to the plane. An adventurous reader can verify, perhaps after reviewing some integration techniques, that the functions $\sin(kx)$ and $\cos(lx)$ are all mutually orthogonal, for any integer values of $k$ and $l$. Similarly, $\sin(kx)$ is orthogonal to $\sin(lx)$, whenever $k \neq l$. When all is said and done, the coefficients $a_k$ and $b_k$ of the best approximation of a given function $f$ can be calculated as follows:

$$
\begin{aligned}
a_0 &= \frac{1}{2\pi}\int_{-\pi}^{\pi} f(x)\, dx, \\
a_k &= \frac{1}{\pi}\int_{-\pi}^{\pi} f(x) \cos x dx \qquad \text{if } k > 0, \\
b_k &= \frac{1}{\pi}\int_{-\pi}^{\pi} f(x) \sin x dx \qquad \text{if } k > 0.
\end{aligned}
$$

It is my sincere hope that the ideas outlined above serve as an invitation and not as intimidation. I made an attempt to present a deep idea on just one page, and the reader might not be ready to make that leap. This is the beginning of an area of mathematics called harmonic analysis, or Fourier analysis. This and many other orthogonal families of functions can be a fascinating subject to be studied. Most of us encounter Fourier analysis on a daily basis. Let us just say, that every time we look at a digital picture in a jpeg format, we are using Fourier analysis to decode it. Every time we listen to music (or to an exciting mathematics lecture!) our brains analyze the sound waves we hear. But this is a whole other story, for another evening or another book.

I wish the reader best of luck in all the mathematics adventures he or she embarks upon. I have always believed that the only way to enjoy mathematics is to understand as much possible and memorize as little as absolutely necessary. I am forgetful too, which is one the reasons this short book was written. I hope that reading it has been helpful and enjoyable, at least some of the time.

## 8.7 Problems

1. Determine if the series $\sum_{n=1}^{\infty} \frac{1}{n^2-10\sqrt{n}}$ converges.
2. Show that the series $\sum_{n=2}^{\infty} \frac{(-1)^n}{n \ln n}$ converges conditionally.
3. Use the ratio test to check that the radius of convergence is $\infty$ for the series in formulas (8.11).
4. Find the integral $\int \frac{\sin x}{x} dx$ by representing the integrand as a series.
5. There is no reason to restrict the study of the power series to real numbers. For any complex number $z \in \mathbb{C}$, define

$$e^z = 1 + z + \frac{z^2}{2!} + \frac{z^3}{3!} + \cdots = \sum_{n=0}^{\infty} \frac{z^n}{n!}.$$

   (a) Write the series for $e^{i\phi}$, where $i^2 = -1$ and $\phi$ is a real number.
   (b) Define $\sin z$ and $\cos z$ for any complex number $z$.
   (c) Calculate $\sin(\phi)$ and $\cos(\phi)$ and combine the answers to write the power series for $\cos(\phi) + i\sin(\phi)$.
   (d) Notice that $e^{i\phi} = \cos(\phi) + i\sin(\phi)$, known as the Euler's formula.
   (e) Put $\phi = \pi$ in the Euler's formula to get

$$e^{i\pi} - 1 = 0.$$

   (f) Order a poster or a T-shirt with this formula!

# Appendix A

# Review of terms

## A.1 Functions

It is good to think of a **function** as a reliable machine $f$, which for the same input $x$ always gives the same output $f(x)$. The **domain** is the set of all acceptable inputs. If no domain is "administratively imposed," one needs to avoid trouble. That is: no zeros in denominators, no negative numbers under square roots, only positive numbers under logarithms and so on. The **range** of a function is the set of all possible values (or outputs). Finding the range can be tricky. It helps to look things like: $x^2 \geq 0$, $-1 \leq \sin x \leq 1$, $e^x > 0$, for every $x$. Finding the $y$-intercept is always easy: if it exists, it is $f(0)$. The **root** of a function means an $x$-intercept. We set $f(x) = 0$ to look for roots. This can be simple or hard, depending on the function.

We are frequently interested in optimizing certain quantities, that is, making them as small or as large as possible. We call $x_0$ a **local minimum** point of $f$ if $f(x) \leq f(x_0)$ in some vicinity of $x_0$, (for $x \in (x_0 - \varepsilon, x_0 + \varepsilon)$, for some $\varepsilon > 0$). We say that $x_0$ is a **global minimum** point of $f$ if $f(x) \leq f(x_0)$ for all $x$ in the domain of the function. Local and global maximum points are defined in the same way, with obvious modifications.

We say that $f$ is **increasing** (or **decreasing**) on some interval $I$, if $f(x_1) < f(x_2)$ for any $x_1$, $x_2 \in I$ such that $x_1 < x_2$ (or the opposite, accordingly).

The **graph** of the function $f$ is the set of points $(x, y)$ which satisfy the equation $y = f(x)$. Not every graph of an equation is a graph of a function. For example, the graph of $x^2 + y^2 = 1$ is a circle, but it does not represent a function. After all, a function is supposed to be "reliable", and for $x = 0$ we would have $f(x) = \pm 1$. In order to represent a function, a set of points on the plane must pass the vertical line test: no vertical line can intersect it more than once.

We say that function $f$ is **continuous** if its graph does not have "breaks": it can be drawn without lifting a pencil. This vague definition is replaced with the proper one in Section 2.3.

A function $f$ is **even** if its graph is symmetric about the $y$-axis, which algebraically means that for every $x$ in the domain, $f(-x) = f(x)$.

A function $f$ is **odd** if its graph is symmetric about the origin, which algebraically means that for every $x$ in the domain, $f(-x) = -f(x)$.

A function $f$ is **periodic** if there exists a constant $p$ such that for every $x$ in the domain, $f(x + p) = f(x)$. The smallest such positive constant is called the **period** of the function.

## A.2 Algebra and graphing

We are interested in how various changes to the algebraic formula defining a function $f$ affect the graph. Let us denote the new function by $g$.

- $g(x) = c \cdot f(x)$. A constant in front of the formula causes a vertical stretch by that constant $c$. This is easy to see, as every point $(x, f(x))$ is replaced by $(x, cf(x))$.

- $g(x) = c + f(x)$. A constant added to the whole function shifts the graph vertically by that constant.

- $g(x) = f(cx)$. A constant in front of the variable alone causes a horizontal *compression* by that constant. This may seem counterintuitive at first, but what happens is that the point $(x, f(x))$ is replaced by $\left(\frac{x}{c}, f(x)\right)$. Another way of looking at it is that the "rubber" $x$-axis gets stretched by a factor of $c$, then the graph of $f$ is drawn and finally the $x$-axis is released back to its original size.

- $g(x) = f(x+c)$. A constant added to the variable alone causes a horizontal shift to the **left** by that constant. Again, we shift the $x$-variable to the right by $c$, then calculate $f(x+c)$. As a result, in order to get the value $f(x)$, we need to start with $x-c$. The point $(x, f(x))$ is replaced by the point $(x-c, f(x))$. (If we really want to shift something to the right, we can temporarily shift the $x$-axis, then draw the graph and then move the $x$-axis back to where it was, pulling the graph to the left.)

- $g(x) = f(x-c)$. Conversely, a constant subtracted from the variable alone causes a horizontal shift to the **right** by that constant.

- $g(x) = -f(x)$. Changing the sign of the value of the function causes the graph to flip about the $x$-axis. All the points of the form $(x, f(x))$ are replaced by $(x, -f(x))$.

- $g(x) = f(-x)$. Changing the sign of the variable causes the graph to flip about the $y$-axis as all points of the form $(x, f(x))$ are replaced by $(-x, f(x))$.

The order of operations is sometimes important. For instance, we can get the graph of $g(x) = 2(x-3)^2 - 5$ by shifting the graph of $f(x) = x^2$ three units to the right, then stretching the graph vertically by 2 and finally shifting it down by 5, in that order. This also tells us immediately that the vertex of the given parabola is at $(3, -5)$.

Bellow is a quick list of basic alebgraic formulas. See [3] or [6] for a more comprehensive treatment.

$$(a+b)^2 = a^2 + 2ab + b^2 \qquad (a+b)^3 = a^3 + 3a^2b + 3ab^2 + b^3$$
$$(a-b)^2 = a^2 - 2ab + b^2 \qquad (a-b)^3 = a^3 - 3a^2b + 3ab^2 - b^3$$
$$a^2 - b^2 = (a+b)(a-b) \qquad a^3 \pm b^3 = (a \pm b)\left(a^2 \pm ab + b^2\right)$$

$ax^2 + bx + c = a\left(x + \frac{b}{2a}\right)^2 - \frac{b^2-4ac}{4a}$, which is why

$ax^2 + bx + c = 0$ for $x = \frac{-b \pm \sqrt{b^2-4ac}}{2a}$, provided that $b^2 - 4ac \geq 0$.

# Appendix B

# Selected Proofs

### Irrationality of $\sqrt{2}$

Suppose that $\sqrt{2} = \frac{m}{n}$, for some $m, n \in \mathbb{Z}$. If both $m$ and $n$ were even numbers, we could we could simplify the fraction. So we can assume that at least one of them is odd. Squaring both sides, we get

$$m^2 = 2n^2.$$

Notice that if $m$ were odd, then $m^2$ would also be odd; a contradiction. So $m$ has to be even and $n$ has to be odd. We can write $m = 2^k p$, where $k \geq 1$ and $p$ is an odd number. Then

$$m^2 = 2^{2k} p^2 = 2n^2$$

Dividing both sides by 2, we obtain

$$2^{2k-1} p^2 = n^2,$$

but this is not possible since the left-hand side is even and the right-hand side is odd.

**Error bound for Taylor polynomials**

Outline of the proof (for $n = 2$ and $x_0 = 0$): Suppose that $|f'''(x)| \leq K_3$ for all $x \in [a, b]$. For simplicity, assume that $x > x_0 = 0$. We want to show that

$$-\frac{K_3}{6}x^3 \leq f(x) - P_2(x) \leq \frac{K_3}{6}x^3.$$

By our assumption we have:

$$-K_3 \leq f'''(t) \leq K_3$$

for all $t \in [a, b]$. This inequality is preserved after integrating over the interval $[0, x]$ :

$$-K_3 x \leq \int_0^x f'''(t)\,dt \leq K_3 x,$$

that is,

$$-K_3 x \leq f''(x) - f''(0) \leq K_3 x.$$

Let us rewrite this again in terms of the variable $t$ and integrate over $[0, x]$ :

$$-\int_0^x K_3 t dt \leq \int_0^x f''(t) - f''(0)\,dt \leq \int_0^x K_3 t dt$$

or

$$-K_3\frac{x^2}{2} \leq f'(x) - f'(0) - f''(0)\,x \leq K_3\frac{x^2}{2}.$$

Finally, we integrate one more time:

$$-\int_0^x K_3\frac{t^2}{2}dt \leq \int_0^x f'(t) - f'(0) - f''(0)\,t dt \leq \int_0^x K_3\frac{t^2}{2}dt$$

or

$$-K_3\frac{x^3}{6} \leq f(x) - f(0) - f'(0)\,x - f''(0)\,\frac{x^2}{2} \leq K_3\frac{x^3}{6},$$

which is exactly what we were trying to show, as $f(x) - f(0) - f'(0)\,x - f''(0)\,\frac{x^2}{2} = f(x) - P_2(x)$.

# Appendix C

# Answers and hints to selected problems

## Chapter 1

1. Center $(-3, 2)$, radius 4.

2. $y = -2x + 10$.

3. a) $m = \frac{f(3)-f(1)}{2}$; b) $m = \frac{f(b)-f(a)}{b-a}$; c) $m = \frac{f(a+h)-f(a)}{h}$.

4. 1E, 2C, 3B, 6A, 8D, 9F, 4,5,7,10 are unmatched.

5. We need to solve the inequality $\frac{(x-3)^2(x-5)}{x^2(x-2)} \geq 0$. Using the method described in Section 1.5 we obtain the set $(\infty, 0) \cup (0, 2) \cup \{3\} \cup [5, \infty)$.

6. Think of $|x-2| + |x-5|$ as the sum of distances from $x$ to 2 and from $x$ to 5. The answer is the interval $(-1, 8)$.

7. Dividing by $x - 1$ we get $x^3 - 2x + 1 = \left(x^2 + x - 1\right)(x - 1)$. The other solutions are $\frac{-1\pm\sqrt{5}}{2}$.

8. $\log_b x$ is the exponent $\alpha$ such that $b^\alpha = x$. The second part also follows from the definition.

11. a) The investment is worth $\left(1 + \frac{r}{m}\right)^{mt}$. If we take $t = 1$ and $r = 1$ (100% APR) we get $\left(1 + \frac{1}{m}\right)^m$. For $m = 1, 12$ and 365 this gives about 2, 2.613, and 2.7146 respectively.

b) The value is $e^{rt}$. This shows that the number $e$ was not invented, but discovered!

12. Notice that $10 = e^{\ln 10}$. Then $10^x = \left(e^{\ln 10}\right)^x = e^{(\ln 10)x}$.

## Chapter 2

1. Try graphing, for example, $f(x) = e^{-x^2/2}$.

2. The stationary points are $-3$ and $0$. (5 is not a stationary point, since $g'(5)$ does not exist. In fact, it is not even a critical point, since it is not in the domain of $g$, but at this stage we do not have the tools to justify that claim.) The function has a local maximum at $x = -3$, by the first derivative test. At $x = 0$, there is no extremum. In fact, it is an inflection point, but again, we need to be patient.

3. a) We estimate from the graph that $p'(1) \cong 0.7$. The tangent line at $(1, 3)$ is given by $y = 0.7(x-1)+3$. b) $p(1.1) \cong 0.7(1.1-1)+3 = 3.07$. c) The function $p$ is increasing whenever $p'$ is positive, that is, on $[0, 2]$ and on $[4, 5]$. It is decreasing on $[2, 4]$. d) The stationary points of $p$ are the $x$-intercepts of $p'$, which are 2 and 4. By the first derivative test, $p$ has a local maximum at 2 and a local minimum at 4. Since the domain is a closed interval $[0, 5]$, the function also has a local minimum at 0 and a local maximum at 5, even though they are not stationary points. e) The function $p$ is concave up when its second derivative $p''$ is positive, which means that the first derivative $p'$ is increasing. We estimate that this happens on the intervals $[0, 0.9]$ and $[3.2, 5]$. f) Possible graphs of $p$ and $p''$ are given below: $p''$

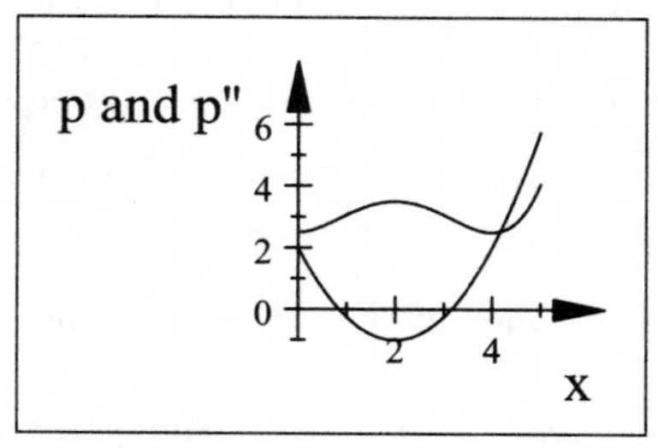


5*. This function is continuous everywhere. The only point at which there could be a problem is $x = 0$. However, $|f(x)| \leq |x|$.

6*. The function is continuous at all irrational values of $x$ and at $x = 0$. This is because, in order to approximate any irrational number by the rationals, their denominators have to grow to infinity. The same thing happens if we approximate 0 by other rationals.

# Chapter 3

1. The least $x^4$ can be is 0, and it happens only when $x = 0$. The first derivative test works, since $f'(x) = 4x^3$ changes sign from negative to positive at 0. However, the second derivative test does not help, as $f''(x) = 12x^2$ and $f''(0) = 0$.

2. a) By the chain rule, $\left(\ln\left(x^2+1\right)\right)' = \frac{2x}{x^2+1}$; by the quotient rule, $\left(\frac{2x}{x^2+1}\right)' = \frac{2\left(x^2+1\right)-2x\cdot 2x}{\left(x^2+1\right)^2} = \frac{2-2x^2}{\left(x^2+1\right)^2}$. b) $(\arctan x)' = \frac{1}{x^2+1}$ and $\left(\frac{1}{x^2+1}\right)' = \frac{-2x}{\left(x^2+1\right)^2}$. c) Be careful: $\ln(\cos(3x))$ is not a product of $\ln x$ and $\cos(3x)$, but a composition, so by the chain rule $(\ln(\cos(3x)))' = \frac{-3\sin(3x)}{\cos(3x)} = -3\tan(3x)$. d) Remember that $e^{x^2} = e^{\left(x^2\right)} \neq (e^x)^2$.

4. If $x = b^y$, then $\ln x = y \ln b$, where $\ln b$ is just a constant. Hence, $\frac{1}{x} = (\ln b)\cdot\frac{dy}{dx}$ and $\frac{dy}{dx} = \frac{1}{x\ln b}$.

6. Keep in mind that $a$ and $b$ are constants. The solution to the initial value problem is $y = 3\cos(2x) + \frac{1}{2}\sin(2x)$.

# Chapter 4

1. Consider $y = \ln(1+2x)^{\frac{1}{x}} = \frac{\ln(1+2x)}{x}$. The limit $\lim_{x\to 0+}\frac{\ln(1+2x)}{x} = \lim_{x\to 0+}\frac{\frac{2}{1+2x}}{1} = 2$, by l'Hôpital's rule. Therefore, $\lim_{x\to 0+}\ln(1+2x)^{\frac{1}{x}} = e^2$. And no, the limit is not 1, even though 1 raised to any power is still one. The trouble is that it is not exactly 1 we are raising to the power of $\frac{1}{x}$.

2. Again, consider $y = \ln(2^x+3^x)^{\frac{1}{x}} = \frac{\ln(2^x+3^x)}{x}$. Using l'Hôpital's rule we obtain $\lim_{x\to\infty}\frac{\ln(2^x+3^x)}{x} = \lim_{x\to\infty}\frac{\frac{1}{2^x+3^x}\cdot(2^x\ln 2+3^x\ln 3)}{1} = \lim_{x\to\infty}\frac{2^x\ln 2+3^x\ln 3}{2^x+3^x} = \ln 3$, as $3^x$ is much bigger than $2^x$ for large $x$. Hence $\lim_{x\to\infty}(2^x+3^x)^{\frac{1}{x}} = e^{\ln 3} = 3$. So 3 wins!

3. The point of this problem is to realize that the values of $\frac{\cos x}{x}$ must stay between $-\frac{1}{x}$ and $\frac{1}{x}$. As $x\to 0$, both $-\frac{1}{x}$ and $\frac{1}{x}$ approach 0. Hence, $\frac{\cos x}{x}$ has no choice but to do the same. Generalize this problem: if two "nice" functions $f(x)$ and $g(x)$ both approach $L$ as $x$ approaches $a$, and the "naughty" function $h(x)$ is always between the two nice ones, $f(x) \leq h(x) \leq g(x)$, then the limit

$\lim_{x\to a} h(x) = L$ as well. This is known as the "**squeeze principle**" or "**sandwich theorem.**"

4. One approach is to express the $x$- and $y$-intercepts as functions of the slope $m$. This makes the area of the triangle a function of $m$. We can now differentiate $A(m)$. Another method is to express the $y$-intercept as a function of the $x$-intercept, which makes the area $A$ a function of the $x$-intercept. The satisfaction of solving this problem will not be taken away from the reader.

5. The cost of the box is $C(x) = 5x^2 + 4\frac{V}{x}$, where $x$ is the side of the base. $C'(x) = 10x - \frac{4V}{x^2}$. This leads to $x = \sqrt[3]{0.4V} \approxeq 0.737\sqrt[3]{V}$ and the height $V/x^2 = \sqrt[3]{6.25V} \approxeq 1.842\sqrt[3]{V}$. In other words, the height should be exactly 2.5 times the side of the base. Having saved all that money on the box, fill it up with fancy chocolates and give it to your favorite person!

7. For $f(x) = \sin x$ we have $P_9(x) = x - \frac{x^3}{3!} + \frac{x^5}{5!} - \frac{x^7}{7!} + \frac{x^9}{9!}$. For $g(x) = \cos x$, $P_9(x) = 1 - \frac{x^2}{2!} + \frac{x^4}{4!} - \frac{x^6}{6!} + \frac{x^8}{8!}$. Note that there is no $x^9$ term this time, as $g^{(9)}(0) = -\sin(0) = 0$.

8. We have $g(x) = x^3$, $g'(x) = 3x^2$, $g''(x) = 6x$ and $g^{(3)}(x) = 6$. Hence $g(1) = 1$, $g'(1) = 3$, $g''(1) = 6$ and $g^{(3)}(1) = 6$. The answer is $P_3(x) = 1 + 3(x-1) + \frac{6}{2}(x-1)^2 + \frac{6}{6}(x-1)^3 = x^3$. We are approximating a cubic polynomial by a cubic polynomial. It had better be the same one.

# Chapter 5

1. $y = \sqrt{4 - x^2}$ is an equation of the circle centered at $(0,0)$, with radius 2. The answer is $\pi$.

2. The function is odd and the areas cancel each other.

3. a) $F(x) = -\int_5^x f(x)\,dx$; b) Represent $G$ as a composition of two functions: $K(x) = x^3$ and $L(x) = \int_1^x \frac{\sin t}{t}dt$. Then use the chain rule; c) Split the interval of integration.

6. For the integral $\int_{-\infty}^{\infty} xdx$ to converge, both integrals $\int_{-\infty}^{0} xdx$ and $\int_0^{\infty} xdx$ would need to converge, but they don't.

7 c) $\frac{e^{-x}}{\sqrt{x}} \leq \frac{1}{\sqrt{x}}$ but this only helps us on the left interval, say $(0,1)$. For larger values of $x$, we need to compare the function with $e^{-x}$.

8. Start with the "midpoint rectangle". On every subinterval imagine a pin at its midpoint. Now turn the top side of the rectangle until it becomes the line tangent to the function. The area of such trapezoid is the same as the

area of the midpoint rectangle. Why?

## Chapter 6

1. a) Put $u = x^2 + 5$; b) by parts: $u = \arctan x$, $dv = dx$;

3. Use $x = \sec t$. When the dust settles, the answer is $\int \frac{1}{x^2\sqrt{x^2-4}}dx = \frac{1}{4x}\sqrt{x^2-4} + C$. Verify it by differentiation!

5. $\int \frac{1}{(x^2+1)^2}dx = \frac{1}{4(x^2+1)}\left(2x - \pi + 2\arctan x - \pi x^2 + 2x^2 \arctan x\right) + C$

## Chapter 7

1. The expression $1 + (f'(x))^2$ is a complete square.

2. Use the fundamental theorem of calculus to write $F'(x) = e^{2x} - 1$.

3. $\int_{-1}^{1}\left(\frac{1}{1+x^2} - \frac{1}{2}x^2\right)dx = \frac{1}{2}\pi - \frac{1}{3}$.

4. Rotate half of a circle to get $V = \frac{4}{3}\pi r^3$.

5. $A = 2\pi \int_a^b f(x)\sqrt{1 + (f'(x))^2}$. The answer is of course $A = 4\pi r^2$.

6. Be careful when taking reciprocal of $\cos t + C$.

## Chapter 8

1. Notice that $n^2 - 10\sqrt{n} > \frac{1}{2}n^2$, for sufficiently large $n$.

2. Use the alternating series test to establish convergence and the integral test to show that $\sum_{n=2}^{\infty}\frac{1}{n\ln n} = \infty$.

4. Start by writing $\frac{\sin x}{x} = 1 - \frac{x^2}{3!} + \frac{x^4}{5!} - \cdots$. Then integrate term by term.

# Appendix D

# Concepts and ideas not to be forgotten

1. The concept of a limit is important for many reasons, including properly defining continuity and derivatives of functions. Plugging in the value of $x$ does not always work! If we get, for example $\frac{0}{0}$, that does not mean that the limit does not exist; it does mean that is more work to be done.

2. The differentiation rules and the derivatives of basic functions such as power, exponential and logarithmic functions, as well as of at least some trigonometric and inverse trigonometric functions (sin, cos, tan, arcsin, arccos and arctan).

3. Implicit differentiation: think of both sides of the equation as functions of $x$. Then differentiate both sides using the known rules. Next, solve the resulting equation for $y'$.

4. Some limits (of the form $\frac{0}{0}$ and $\frac{\infty}{\infty}$) can be calculated using l'Hôpital's rule: $\lim_{x\to a}\frac{f(x)}{g(x)} = \lim_{x\to a}\frac{f'(x)}{g'(x)}$. There is no rule like that for the product, which needs to rewritten as a quotient before attempting l'Hôpital's rule: $f \cdot g = \frac{f}{1/g}$. Also, sometimes it is helpful to find the limit of the logarithm of the function first: $f(x) \to L$, whenever $\ln(f(x)) \to \ln L$.

5. In optimization problems the most difficult part is frequently the set-up. It is important not to mix up the objective function with the constraint.

6. Related rates: do not plug in numbers too soon. Keep the variables as variables until after differentiating. Put the numbers in at the end.

7. Differentiable functions can be locally approximated by the tangent line - the first order approximation. Linear approximations of a function at various points are used extensively in numerical methods. For example, Newton's method is used for algebraic equations and Euler's method for differential equations. For a better approximation of a function we can use Taylor Polynomials.

8. The intermediate value, the extreme value and the mean value theorems are examples of existence theorems. It is good to know what they say.

9. The definite integral is the signed area below the graph of a function over a given interval. It is formally defined as a limit of Riemann sums. The value of the definite integral can be approximated numerically using rectangles or trapezoids instead of the actual function.

10. The fundamental theorem of calculus. The derivative of the area function is the integrand:

$$\tfrac{d}{dx}\int_a^x f(t)\,dt = f(x), \quad \text{that is,} \quad \int_a^b f(x)\,dx = F(b) - F(a),$$

    where $F$ is any antiderivative of $f$.

11. Finding antiderivatives is a lot harder than finding derivatives. It takes some effort to learn how to differentiate products and quotients, but at least there are rules. Knowing that there are elaborate techniques for integrating products, one should never attempt to integrate products or quotients of functions "one piece at a time."

12. The chain rule for differentiation corresponds to the method of substitution for integration. It does not work well for every integral and the substituting variable (say $u$) must be chosen carefully. Make sure that all the remains of the old variable are gone before attempting to integrate with respect to the new variable. Also, do not forget that $dx$ does not automatically become $du$, but that $du = u'\,dx = \frac{du}{dx}dx$.

13. Integration by parts, (the poor man's product rule): $\int udv = uv - \int vdu$.

    The goal is to find two functions $u$ and $v$, such that after applying integration by parts the new integral is easier to find than the original. The trick is to choose the functions wisely. Another strategy is to come back to the original problem, usually after integrating by parts twice, with a coefficient different from 1. Solving the resulting equation for the unknown integral gives the answer.

14. Partial fractions are useful for integrating rational expressions. One should know how to divide polynomials and how to rewrite a proper rational expression as a sum of partial fractions.

15. Trigonometric antiderivatives. These are interesting, not only for their own sake, but because they are useful in integrating expressions involving $\sqrt{a^2 - x^2}$, $\sqrt{x^2 - a^2}$ and $\sqrt{x^2 + a^2}$. For the purposes of integration, some of the most frequently used trigonometric identities include the Pythagorean, double angle and half angle identities.

16. Integration can be a complicated process and even the best of us make silly mistakes once in a while. However, whenever feasible, check your antiderivative by differentiation to avoid giving obviously wrong answers!

17. Some applications of the integral include

    (a) The arc length (from Pythagorean theorem) $L = \int_a^b \sqrt{1 + (f'(x))^2} dx$.

    (b) The area between two curves $A = \int_a^b Bigger(t) - Smaller(t)\, dt$.

    When the variable is $x$, then "Bigger" means higher up, and when the variable is $y$, then "Bigger" means more to the right. One of the difficulties is finding the intersection point of the curves.

    (c) Volumes of solids by cross-sections $V = \int_a^b A(x)\, dx$,

    in particular the method of discs or washers when finding the volume of a solid of revolution. Cylindrical shells can also be useful.

18. Integrals can be improper because one or both limits of integration are $\infty$, or because of a vertical asymptote of the integrand somewhere in the interval of integration.

19. To show convergence of an improper integral, one can find a friendly function greater than the one in question, whose integral converges. On the other hand, to show divergence of an improper integral, find a friendly function whose integral diverges. Friendly functions are those whose convergence or divergence is known or easily determined by direct integration, such as $\int_1^\infty \frac{1}{x^p} dx$ or $\int_0^\infty e^{-x} dx$.
20. We use infinite series to add infinitely many numbers. The sum of an infinite series is the limit of the sequence of partial sums. A series $\sum a_n$ and a sequence $a_n$ and should not be mixed up. In particular, for the series $\sum a_n$ to converge to anything, the sequence $a_n$ must converge to 0. This condition is necessary, but not sufficient: a good example to keep in mind is the harmonic series $\sum_{n=1}^\infty \frac{1}{n} = \infty$, even though $\frac{1}{n} \to 0$ as $n \to \infty$.
21. There are a few series whose sum we can express in a closed form. These include various kinds of telescoping series (a curiosity of sorts) and the geometric series (very important), for which we have
$$\sum_{k=0}^{\infty} ar^k = \frac{a}{1-r}, \qquad \text{if } |r| < 1.$$
22. Various methods are used to determine convergence of series. These include comparing a given series with a "friendly" one (see #19), the integral test (useful, for example, for the $p$-series) and the ratio test, which measures how closely a given series resembles the geometric series.
23. We say that a series $\sum a_n$ is convergent absolutely if the series of absolute values $\sum |a_n|$ is convergent. The terms of an absolutely convergent series can be reshuffled without affecting the value of its sum.
24. Power series: $\sum_{k=0}^\infty a_k (x - x_0)^k$ . Their convergence depends on the value of $x$. The set of those values of $x$, for which the power series converges is a symmetric interval centered at $x_0$. The radius of convergence (one half the length of the interval of convergence) is usually found using the ratio test. Convergence at the endpoints of the interval of convergence should be examined separately.
25. Inside the radius of convergence we may differentiate and integrate functions written as a power series, "term by term."

# Bibliography

[1] Marlow Anderson, Victor Katz, Robin Wilson (eds.), *Sherlock Holmes in Babylon and Other Tales of Mathematical History,* The Mathematical Association of America, 2004

[2] Eric Temple Bell, *Mathematics: Queen and Servant of Science*, The Mathematical Association of America, 1987

[3] I. M. Gelfand, E. G. Glagoleva, E. E. Shnol, *Functions and Graphs*, Dover Publications, 2002

[4] I. M. Gelfand, Mark Saul, *Trigonometry*, Birkhäuser Boston, 2001

[5] http://www.pbs.org/wgbh/nova/archimedes/

[6] George Simmons, *Precalculus Mathematics in a Nutshell*, Barnes and Noble Books, 1997

# Index